Praise for *Move UP*

'Choices, movements, evolution, requirements for survival of the species. Why do some of us strive for more? How are culture, surroundings and education helping some societies move up more than others? These are some of the questions the authors engage by studying the paradigm between social and biological sciences. *Move UP* portrays the cultural and biological dimensions behind the desire of humans to ascend socially, intellectually or economically. Surprisingly being logical about our biology seems to be the key!'

Mario J. Molina, President of the Mario Molina Center on Energy and Environment, and 1995 Nobel Prizewinner for chemistry

'*Move UP* is a splendid book, totally engaging from start to finish. It showcases the human drive to strive for betterment within a complex matrix of our biology and culture. It challenges the reader to think about fresh ideas about ways to ascend, as well as highlighting the impediments that must be overcome to progress toward improvement. *Move UP* is by far the best book I've read this year.'

David M. Buss, author of *Evolutionary Psychology: The New Science of the Mind* and *The Evolution of Desire: Strategies of Human Mating*

'*Move UP* provides a provocative and entertaining look at interactions between culture and biology that impact the progress of societies. The authors raise big questions and challenge the reader to think about them in novel ways.'

Daniel L. Schacter, William R. Kenan, Jr. Professor of Psychology, Harvard University, and author of the *The Seven Sins of Memory: How the Mind Forgets and Remembers*

'*Move UP* considers a critical question in our globalized world – what sorts of countries foster social mobility in their citizens? This is a question long pondered by social scientists, but Rapaille and Roemer bring a fresh perspective to the question, viewing the subject from the standpoints of evolutionary biology, anthropology and zoology. The result is a superb book – provocative, smart, fun to read and very important. I recommend it highly.'

Robert Sapolsky, neuroendocrinologist and author of *Why Zebra's Don't Get Ulcers* and *Monkeyluv: And Other Essays on Our Lives as Animals*

'You have just been elected head of your country. You want to lead your people to new heights of happiness, prosperity, security, and freedom never enjoyed before. What should you do? The first thing you should do is read *Move UP* by Clotaire Rapaille and Andrés Roemer, and follow their data-driven recommendations for determining how best to achieve your goals using science, technology, and the wisdom of the greatest thinkers in history revealed in this remarkable book. *Move UP* is not utopian; it outlines a realistic plan for how more people in more places more of the time can lead more fulfilling and successful lives anywhere in the world.'

Michael Shermer. founder and editor of *Skeptic* magazine, and author of *The Believing Brain: From Ghosts and Gods to Politics and Conspiracies – How We Construct Beliefs and Reinforce Them as Truths* and *How We Believe: The Search for God in an Age of Science*

'Andrés Roemer has been a leader in bringing important scientific ideas to public attention, as well as promoting human rights and effective democracy; Clotaire Rapaille has been a leader in the psychology of generating effective marketing strategies. Now, in *Move UP*, Drs Roemer and Rapaille combine their talents as scientists, communicators, motivators and activists – with a little help from a uniquely diverse cast including Charles Darwin, Albert Camus, Sigmund Freud, Dr Seuss and Napoleon Bonaparte – to show how individuals as well as societies can move: which way? UP, of course! If you value 'survival, sex, security and success' (and who doesn't?) then get ready for a potentially life-altering trip!'

David P. Barash, professor of psychology
at the University of Washington and author of
Homo Mysterious: Evolutionary Puzzles of Human Nature

'Rapaille and Roemer fix their gazes on a question of great importance and intimidating complexity: how can we prosper? With inviting style, they bring to bear an array of ideas and an abundance of evidence, looking closely at the issue from a host of different vantage points. Ranging widely, the authors will delight, provoke, and very possibly inspire readers who want to know how nations can move UP.'

Robert Kurzban, associate professor of psychology at the
University of Pennsylvania, and editor-in-chief of
Evolution and Human Behavior

'I think this book is terrific . . . In *Move UP*, Clotaire Rapaille and Andrés Roemer have written a book that is engaging, stimulating, and challenges us to think in new ways. Though it is theoretically broad and ambitious, it is full of practical suggestions that can make life better. It will change the way you look at almost everything.'

Barry Schwartz, author of *The Paradox of Choice: Why More is Less* and *Practical Wisdom*

'What are the parallels between brains and cultures? Running the gamut from stimulating to provocative, heady to poignant, this book gives you plenty to think about for moving lives and societies in the only useful direction.'

David Eagleman, neuroscientist and author of *Sum: Forty Tales from the Afterlives* and *Incognito: The Secret Lives of the Brain*

'I love the ideas put forth by my good friends, Roemer and Rapaille! The Culture Code of your home country is absolutely critical for a successful future . . . Culture matters, *a lot*! The good news is we can change, one by one, we can change our mindset. It's up to us. You must decide to move UP, and maybe away from your home. Read the book and find out why and how.'

Ricardo Salinas Pliego, founder of Grupo Salinas

'When does a nation's culture hold them back, and when does it fuel their progress? It is extremely hard to answer this question comprehensively enough to help guide policy change – whether this be economic policy, social policy or even foreign policy. Roemer and Rapaille offer a cogent and coherent network of guidelines, describing not only what has worked and not worked in the past but also a set of persuasive arguments and theories as to why. Those who would tend to challenge these guidelines on the basis that they are too 'Western', or in some other way parochial, would do well to ask themselves how long other paradigms need to fail before they should be abandoned.'

Aubrey de Grey, Chief Science Officer of the SENS Research Foundation, and co-author of *Ending Aging: The Rejuvenation Breakthroughs That Could Reverse Human Aging in Our Lifetime*

'Clotaire Rapaille and Andrés Roemer are great storytellers and natural provocateurs, and *Move UP* is a treat – a clearly written and very creative exploration of the conditions that lead to happiness, freedom, and flourishing.

Paul Bloom, professor of psychology at Yale University, and author of *Just Babies*

'This is the best book I have ever read about sex, success, survival, security, and the reptilian brain. Roemer and Rapaille have done an amazing job explaining difficult concepts about the human condition and life in the modern world in a fascinating, humorous, entertaining and elucidating way. I recommend this book very highly to anyone interested in learning about what life is about – and about what it can be.'

Amir D. Aczel, mathematics enthusiast and author of *Fermat's Last Theorem*

'*Move UP* is a tour de force presenting a muscular new formula for individuals and countries to assess – and perhaps fix – their barriers to upward mobility, creativity and talent. A deep and entertaining read.'

Louann Brizendine, author of
The Female Brain and *The Male Brain*

'How can we explain why some societies are successful while others are not? How can we engineer societies that best satisfy fundamental human needs? Over two thousand years ago, Plato compared the well-functioning society to a well-functioning soul – a soul with three parts all working in harmony. In this highly engaging and accessible book, Andrés Roemer and Clotaire Rapaille update Plato's project using Maclean's model of three brain systems to draw out what it is that makes the difference between societies that progress – that 'move up' – and those that stagnate and fail. This is a stimulating and thought-provoking book that is full of practical wisdom. I greatly enjoyed reading it, and I think that you will to.'

David Livingstone Smith, psychologist and
philosopher, and author of *Why We Lie:*
The Evolutionary Roots of Deception
and the Unconscious Mind

'A revolutionary take on cultural mobility, elegantly composed yet easy to understand. *Move UP* serves as a whirlwind tour through history and science, exploring what drives success for individuals and societies. Bravo to Rapaille and Roemer for introducing a novel way to quantify the human condition. This interdisciplinary and thought-provoking book may reshape the way readers think about our world.'

Sheril Kirshenbaum, author of *The Science of Kissing*
and co-author, with Chris Mooney, of *Unscientific*
America: How Science Illiteracy Threatens Our Future

Move UP

Why Some Cultures Advance and Others Don't

CLOTAIRE RAPAILLE
AND ANDRÉS ROEMER

ALLEN LANE
an imprint of
PENGUIN BOOKS

ALLEN LANE

UK | USA | Canada | Ireland | Australia
India | New Zealand | South Africa

Allen Lane is part of the Penguin Random House group of companies
whose addresses can be found at global.penguinrandomhouse.com.

First published in Spanish by Taurus Publishing, Spain 2013
This English edition published 2015
001

Set in 12/14.75 pt Dante MT Std
Typeset by Jouve (UK), Milton Keynes
Printed in Great Britain by Clays Ltd, St Ives plc

A CIP catalogue record for this book is available from the British Library

ISBN: 978-0-241-18699-2

www.greenpenguin.co.uk

Dr Clotaire Rapaille

I would like to dedicate this book to all the men and women who helped mankind move up. From the first human being who stood up on two feet to Neil Armstrong's first step on the moon. We are just at the beginning of the journey.

I want to thank the few who have inspired me throughout my life. As a young teenager, it was Voltaire, Descartes, Thomas Jefferson and Alexis de Tocqueville. Then, as I grew up, Levi Strauss, Henri Laborit, Ruth Benedict and Edward T. Hall.

But most of all, there is a baby girl who was born recently in one of the worst places in Africa. This book is for her. She is the future of this planet and I am sure that one day she will teach the new generation how to move up.

Dr Andrés Roemer

To my immortal beloved ones:
Alejandro, David, Valeria;
Fanny, Oscar;
Nadia;
and Pamela.
I wish you – and future generations – a world free of shackles and full of enlightenment

In memory of Christopher Hitchens, 1949–2011.

CONTENTS

ACKNOWLEDGEMENTS

Dr Clotaire Rapaille

My first acknowledgement is to an American GI driving a jeep in France several weeks after D-Day. He gave me some chocolate and chewing gum, and he also provided my first imprint of 'liberator'. He made me want to become an American.

My second acknowledgement is to Lee Kwan Yew, the founding father of modern Singapore, who gave me hope for mankind. With no oil, no gas, no land, no resources, starting from scratch in 1965, he created one of the most successful cultures on this planet. His accomplishment has become a model for China and should be a model for the next generations.

I'd like to thank my mother, who in very difficult times always told me, 'It doesn't matter what you want to do, you will always succeed.' She gave me the confidence to know that, when you want to move up, you can.

I have been inspired by many great people throughout my life, too many to mention here. But Dr Roemer, my partner in crime, is definitely one of them. His total loyalty and confidence in our ability to create this index has made this book possible.

Dr Andrés Roemer

Two institutions have supported this book from beginning to the end. First, the University of California at Berkeley, particularly the School of Law and its faculty have guided me for the past two years while I have been a Senior Research Fellow. I would be remiss if I didn't mention in particular Robert Cooter, my professor and mentor. The second is Poder Cívico, which, along with its project La Ciudad de las Ideas, has become my new baby to raise.

I'd also like to express special gratitude to my friend Ricardo Salinas Pliego, who believed in this project (probably because he himself is always moving up), and who has generously taught me many of the ideas found in this book. Also to José Antonio Meade, a Mexican mensch and a visionary with whom in 2012 I produced the brainstorming session Rethinking G20: Designing the Future, in Cabos, Mexico.

It's impossible not to mention Dr Rapaille. He introduced me to the world of culture codes, and is my colleague, my life teacher and a genuine archetype of genius. But above all, a great friend who reinvents himself and moves up along with those who are fortunate enough to learn from him.

And of course I'd like to thank my parents, Fanny Slomianski and Oscar Roemer, who taught me that to move up is about 'going from failure to failure with great enthusiasm'. My greatest thanks go to my children, Alejandro, David and Valeria, who have been my greatest intellectual challenge, my raison d'être and, most importantly, my inspiration to never stop moving.

Both Authors

For the quantitative part of our index, we must recognize the research team from the School of Government and Public Policy (EGAP) at Monterrey Institute of Technology and Higher Education (ITESM), State of Mexico Campus, who helped in the research, methodology and data compilation for the R^2 Mobility Index. Special thanks to Marisol Serna, who helped us gather information and data for the index. The Espinosa Yglesias Research Center (CEEY) also prepared several reports and articles on social mobility in Mexico that helped us understand mobility in the context of today's economic situation. It's also worth mentioning that our index would not be possible without the thorough and high quality research conducted by the United Nations, *The Economist* Intelligence Unit, the Heritage Foundation and the *Wall Street Journal*, the World Bank, the International Monetary Fund and the World Economic Forum, among other universities and research institutions.

We'd also like to thank Pamela Cortes, Eidi Cruz, Fernando Meneses and Osseily Hanna for their comprehensive revision of the final text. We're grateful to our English publishers, Penguin Press, and particularly to Cecilia Stein who has always believed in this book. And to Jesse Steele and Cármen Zárate who have supported us throughout this project.

We are very grateful to our chief editor for the English version, Francisco Brito Gómez, for coordinating the research team, editing and interpreting our ideas. Emily Loose provided us with very useful comments, and we are grateful to her. As part of the research team, Veronica M. French and Jorge Arroyo did a fantastic job helping us put together these ideas.

For the qualitative part of our index, we must acknowledge the hundreds of people from around the world who were involved in our studies, participating in focus groups and qualitative studies in more than seventy-one countries. Thanks forever to the more than 2,730 people who were part of this project.

Naturally, we are accountable for any errors and omissions in this work.

Preface: The Book Most Travelled

It is trite but true that every book is like a child's birth: some have a simple and fast delivery, others not so much. This book took years to fully develop.

We first thought of writing it in 2009, during the World Economic Forum (WEF) in Davos, Switzerland. As we celebrated Clotaire's wife Missy's birthday, we raised our glasses to the idea of a new social mobility index, an indicator that would not only reflect what a country does in terms of mobility and prosperity for its citizens, but would also incorporate variables that can truly promote social mobility among people and nations. These are the bio-cultural variables: the C^2 and biological variables that are discussed in this book. Our vision kept cropping up throughout Missy's birthday celebration, and the WEF conferences failed to touch on the topic, leaving us with unanswered questions: how can we move up? What moves us, and what *allows* us to move?

We continued to discuss the topic after our stay in Switzerland, sharing our experiences and discoveries, and each time finding something new. From Paris to Mexico, Mexico to Miami, Miami to New York, and again in Paris; our quest for answers became an obsession and we kept digging. (See the Bibliography for a full list of the books where many of our theories and ideas in *Move UP* have come from.)

While working on this book, we've both travelled around the world and talked with a wide variety of people from different backgrounds. During our travels we noticed a powerful archetype that exists in all cultures, and serves as a trustworthy indicator of the evolution of a given culture. This archetype is encapsulated in the phrase 'move up'. It is a multi-faceted notion and the most definitive indicator of well-being. In most cultures, a strong sense of social, financial and physical growth is closely linked to the sense of mobility and vitality that a given culture has. Cultures that understand what it means to move up have a firm place on the global stage.

Indeed, this is a book about the biggest question we can ask: 'Who are we?' When we ask this, we are asking: 'How can we live a more meaningful life?' or 'How can we move up?' This is a critically important question, one that leads many people to read self-help books, to diet, enter therapy, join religious groups, to exercise, to meditate or to practise yoga. In this book we will show that those individuals, cultures, societies and countries that understand movement most are much more likely to express themselves in a creative way: innovating, transforming their environment, and making a positive impact on society.

This is not an academic text, rather a book for the curious. It's not about the economic growth of countries, or a self-help book that will give you tricks for succeeding. We're not trying to measure social mobility, since many capable sociologists, economists and anthropologists continually attempt to do so. *Move UP* is about uncovering the reasons *why* some cultures move more than others.

What *Move UP* proposes is a comprehensive answer to the question of *why*: why do some people have the opportunity to move in the directions they want to, while others don't? Why are some societies more mobile than others? Why do people move up and others seem to stay static? These are not easy questions, and therefore the answers cannot be easy either. So let's get moving.

The question of what enables some countries to provide their people with increasing prosperity while others fail to do so is a long-standing point of debate, periodically erupting into controversy – and driving sales of many bestselling books. One camp of scholars has focused on the pernicious effects of colonialism and how it led to a dependency of poor nations on rich ones. Other leading arguments have focused on the happenstance of geography, in particular the climate and natural resources countries are endowed with. In his bestselling *The Wealth and Poverty of Nations*, published in 1998, economist and historian David Landes quotes John Kenneth Galbraith on this: '[If] one marks off a belt a couple of thousand miles in width encircling the earth at the equator one finds within it no developed countries.' Landes offered a number of explanations as to why hotter climates hindered prosperity, and led to cultural practices and economic and political institutions that have compounded this, from siestas that limit daily productivity to the fact that slaves were mostly sourced from these areas (due to manual labour being so onerous in hotter climates) and to hotter zones being less fertile for agriculture while also being breeding grounds for so many lethal diseases. At the same time he highlighted the industriousness of the more temperate nations of Europe as the key to their success. He was praised by some critics for putting the focus on the culture and institutions within the impoverished nations and pilloried by others for Eurocentrism and failing to attend to the rise of the Asian economies. Around the same time, Jared Diamond, in his 1997 bestseller and Pulitzer prize winner *Guns, Germs and Steel*, offered his own nuanced argument about how environmental factors drove the differences in innovation of technologies and institutions that accounted for the success of the west and the booming Asian economies. Leading development specialist Jeffrey Sachs has also stressed the role of natural resources and climate in his assessment of underdevelopment, particularly in Africa, arguing in his 2005 bestseller *The End of Poverty* that 'Africa's governance is poor because Africa is poor.' He has stressed the importance of foreign aid, asserting that there is no reason the poorer countries can't catch up if they are given adequate assistance. While the UN appointed him the director of its Millennium Project and tasked him with overseeing the crafting of its Millennium Development Goals, and a great deal of money and government support has been devoted to initiatives guided by him, he's also been sharply criticized by some who argue that the kind of poverty-eradication projects he advocates have actually left countries worse off and have deepened their dependency.[1]

Among the most recent entries into the fray, one of the most compelling arguments is that of Daron Acemoglu and James Robinson in their bestselling book *Why Nations Fail*, who, while conceding that nations with temperate climates have had an advantage over those in the tropical zone, argue, 'World inequality . . . cannot be explained by climate or diseases, or any version of the geography hypothesis.'[2] They cite the dramatic differences in development on either side of the Mexican–US border, between North and South Korea, and between East and West Germany before integration as cases that refute those explanations. Building on a rich foundation of work in the field of institutional economics, they assert that the political and economic institutions a country creates are the deciding factor, pointing in particular to the establishment of relatively egalitarian property rights, constitutional democracy and relatively equitable access to capital as the keys to the breakaway success of the western nations.

We are grateful for all of this work: all insights into how the nations of the world can better attack the problem of persistent poverty and provide a higher standard of well-being for their people are vital. In this book, we seek not to criticize these efforts, or to choose sides, but to build on this existing work. We believe that there are important limitations to all of these approaches in explaining the prosperity of nations, and especially in evaluating how well they foster the well-being of their citizens.

When we met in Davos we soon began to discuss how the focus on the size and growth of GDP as the key indicator of the success of nations has obscured the importance of a wide range of other contributors to the quality of life. For one thing, it has tended to downplay some of the festering problems in the GDP-rich nations, such as the many communities wracked by extreme poverty in the US, and the persistence of corruption and extreme poverty in the emerging nations. And it has also taken the focus away from certain cultural practices and beliefs that either contribute to prosperity and well-being or weigh a country's people down.

We realized that we had complementary expertise – cultural analysis and public policy training – that would allow us to collaborate on conducting research into a better way of evaluating national success and to craft a new, more robust way of understanding how well a country is fostering well-being for its people, and how well it is likely to do so in the future.[3] We wanted to offer a more human-centric, holistic way of analysing the problems, and prospects, of countries in order to help craft more effective ways to bring about change. Whatever your opinion of the current predominant approaches to global development, it is difficult to claim that the rate of improvement is at all adequate. In many countries conditions are actually worsening, and in many of both the developed and developing countries deep problems persist.

We are not trying to measure social mobility, since many capable economists, sociologists and anthropologists are already looking into this. *Move UP* is about uncovering the reasons why some cultures move faster than others. Why Singapore, which started with nothing in 1965 became one of the safest states and richest countries in the world, whilst France, which can claim a very rich past, is seeing both its crime rate and national debt go up and its young people moving out of the country in search of better opportunities elsewhere. What is it that the Singaporeans are doing right and the French are doing wrong?

For over forty years now, since we were teenagers, we have been struck by all the rhetoric surrounding the supposed 'decline' of the United States: 'This is it,' the pundits claim, 'America is going down – it's over.' Yet, time and again, America has come back stronger than before. All the intellectual explanations and logical analyses were unsatisfactory. No economic model could adequately explain this trend, and statistics were misleading. Take average income as a benchmark, for instance. It doesn't mean much. If ten people make $1,000, then the average income is $1,000 per person. Yet if one of those people makes $100,000, suddenly the average goes up from $1,000 per person to $109,000 ÷ 10 = $10,900 per person. So the fact that one person is now making $100,000 changes the average completely, yet the majority are still making the same amount of money. The result, looking at the statistics alone, tells us that everybody is more than ten times wealthier than they are in reality.

When you read that the average Singaporean makes $60,000 USD a year and that the average French person makes $25,000

USD a year, what does it mean? When the French president says that he hates the rich and decides to tax them at a 75 per cent rate, what is the result? The result is that these wealthy people will leave, and thus the average goes down. Does this mean that the majority of people are making less money? After studying numbers and statistical data around the world, we became aware that there might be a better explanation and that all of the models available were not giving us clear insights into why these things were happening. Most of the time these analyses are performed by economists and statisticians who count numbers and compare them, like bean counters.

Singapore has a surplus of $55 billion a year whereas France has a deficit of around that same number. What are these numbers really telling you? If you realize that there are only 6 million people living in Singapore compared to 60 million people in France, it's clear that the performance of the individual carries much more weight in Singapore. The Singaporean surplus becomes much more significant when you take into account that it has been reached by only 6 million people, with no oil, no gas, no natural resources and very little land. What are they doing right, and how are they able to maintain an unemployment rate of under 2 per cent when the French are at 11 per cent and the Spanish at 25 per cent? These numbers are no accident and are far from being random. There is something about the Singaporean culture that keeps their unemployment rate this low: is it because it is safe and clean? In France, by contrast, you have the development of a nervous middle class.

When Singapore was separated from Malaysia in 1964, its inhabitants had nothing, no land, natural resources, economy or agriculture. How did they create a culture that is so successful today? Lee Kuan Yew told his people, 'One word: clean. We are going to start with clean. Everything is going to be clean, your clothes, home, and bodies.'[4] Just by focusing on one simple dimension within a culture, society changed, becoming more organized and disciplined, and ultimately helping individuals move UP.[5]

The movement of populations is also a good indicator of which culture is moving up and which is moving down. This is because people vote with their feet: when they cannot move up, they leave. If you look at the walls countries build, they are created either to keep people in or to keep people out. After World War II, the purpose of the Berlin Wall was to stop people from leaving. In the US, the wall between the US and Mexico stops people from coming in. This is a good indicator of how competitive a country really is.

We have travelled the world for several decades trying to decode cultures and asking ourselves challenging questions, such as: why are most of the Catholic countries poor and most of the Protestant ones rich? Why do the Japanese and Germans make good soldiers and know how to efficiently manufacture cars, whereas the Argentinians do neither but are successful at exporting meat? Are there 'car cultures' and 'meat cultures' or is something else going on?

We are certainly not the first to attempt to propose a more human-centric evaluation of how nations provide for well-being. One interesting alternative measure to GDP that has sought to focus on quality of life in addition to economic prosperity, and has garnered a good deal of attention, is the concept of Gross National Happiness (GNH), pioneered by the government of tiny Bhutan. This is a valuable concept, but we don't believe that measuring levels of happiness truly gets to the heart of what accounts for that happiness; or that happiness is necessarily the best measure of whether a country is enabling its people to thrive. Indeed, as the father of happiness studies, Martin Seligman, has argued, a good deal of innovation and drive can arise out of discontent.

As we continued to discuss the issue and to think about all the factors that contribute to a nation's ability to foster the well-being of its individual members, and the more holistic prosperity of the society as a whole, we zeroed in on an underlying, driving force that we argue is the key determinant of both prosperity and the level and distribution of well-being in a country: what we call 'moving up'. What we mean by moving up is, in essence, the degree to which a country enables its people to evolve both socially and economically, and also, crucially, to become happier and feel more purposeful. In proposing moving up as the goal to strive for, we want to focus on how much opportunity there is for growth in social, financial, physical and emotional well-being within a country. We mean moving up not only in terms of advancing up the educational and economic ladders, but moving up in the understanding of issues and seeking of solutions within the country, moving up in the evolution of cultural practices and beliefs, and in the movement towards better institutions and policies that foster well-being. So mobility is increased not only by better educational opportunities, better access to quality medical care and a healthy diet, the rooting out of corruption in business and government, the free exchange of ideas in an open national media, and the fostering of innovation in all fields of endeavour. Also contributing to mobility is whether a country fosters the abandonment of cultural practices that are harmful to all or some of its people and are limiting their opportunities for personal growth.

We became aware that it is important to move, but this movement needs to have a direction. Mothers know that for their child to grow and move up, they have to be clean, healthy, well fed, loved and protected. Moving up is the natural direction or *sens*, which in French means both 'direction' and 'meaning'. So we looked at cultures and nations around the world, not just in terms of statistics such as GDP, but in terms of how they facilitate or prevent people from moving up. It became obvious that priority should be given to biology, or the natural course of life (what we dub 'bio-logic').

If life is movement and children have to grow, movement and growth are important. Some cultures keep you prisoner by restraining your movement and punishing success and growth. Even some basic biological stages of life are repressed in some cultures and celebrated in others. We studied menstruation for the brand Always and found that, in some cultures, when a girl experiences her first period she is made to feel ashamed. She is sometimes ostracized during that particular week or so. In other cultures, however, the event is celebrated because it signifies the transition from girlhood into womanhood. As part of this celebration, the girl receives gifts and praise. She is made to feel proud.

Whilst studying seduction for L'Oréal, it became clear that 'seduction' in Germany is not the same as 'seduction' in Italy. Studying cleanliness for Procter & Gamble, it became obvious that the Germans are very clean, whereas the Indians and the Chinese are not so clean. If you go into a Japanese bathroom, it's as if you're walking into a spaceship. It became clear that cleanliness, seduction and food were related to the private dimension of life that we call the reptilian (in reference to the reptilian brain), which deals mainly with survival and reproduction. If a culture tells you that you can never have sex, then there will be no children and this is the end of the species. So success and growth are related to survival and reproduction, and to how the culture satisfies basic biological needs. We became aware that some cultures help you to survive and reproduce, whereas others make this difficult.

We found profound differences between the way Americans work and perceive being paid for their services – take for instance a typical example of a child being paid to wash the family car – and the way the French reject money and celebrate *l'acte gratuit*. In order to move, to grow and to prosper you need a culture that gives you both the opportunity to progress and a reward structure to back it up.

Moving up became our obsession: how, when and where can people move up. To study this, we needed a new way of looking at cultures and their relationship with biology. We went around the world studying health, hygiene, medication, medicine, food and nutrition for various pharmaceutical companies. We studied concepts of sense, sensuality and seduction for Firmenich, L'Oréal and Estée Lauder. We explored sex and sexuality for Trojan and Johnson & Johnson. From India to Brazil, China to Mexico and Russia to Australia we kept looking at the relationship between biology and culture.

We think that to foster well-being for the world's people – both in the developed and developing worlds – the global conversation has to shift to the goal of fostering mobility. In order to start this conversation, we realized that we would have to create a way to assess and compare the degree to which different countries foster mobility. That meant identifying a robust set of key factors that contribute to it. In doing so, we agreed that no good measure could be created that didn't account for two vital factors that are most often given short shrift or ignored entirely: the role of biology and the role of culture.

Let us consider these in turn.

Biological imperatives are universal. No matter whether we are French or Brazilian, Inuit or Maori, Jewish, Muslim or Buddhist, we are all driven by the same biological forces. This has been established beyond dispute. And these forces propel us not only to survive and multiply, but to thrive. Or, as we like to put it, to move up. We are in effect biologically programmed not only to compete for the resources we need for survival and to pass on our genes by reproducing, but to experience pleasure, to learn, to establish strong relationships, to figure out ways to cooperate, to express ourselves, to enjoy pleasure. Abraham Maslow asserted as much with his theory of self-actualization sixty years ago, and a great deal of work in the years since has illuminated the biological roots of this drive to thrive, as well as the many ways in which our biological impulses can turn against us and our society if they are repressed.

Classical economics – and all of the prevailing theories of national development based on it – presupposes that rational decision-making explains the fundamental workings of economies. But a great deal of work over the course of the last few decades has revealed just how limited that view of human behaviour is. The field of evolutionary psychology, aided by discoveries in neuroscience, has offered fascinating and powerful explanations of how our evolution has shaped our behaviour, from our propensity to go to war, to the motivations behind murder and rape, our fierce loyalty to our blood relations and larger 'tribes', the biological logic of altruism and community building, our drive for status and the need to feel respected. Cognitive psychology has complemented this work by providing a wealth of insight into how irrational our thinking so often is due to glitches built into our minds. Daniel Kahneman's masterful *Thinking, Fast and Slow*[6] introduced the wide range of these 'cognitive illusions' that lead us into bad judgements, from being overconfident to jumping to biased conclusions, and making badly flawed assessments of risk. This work has shown that one reason the reptilian so often overpowers the thinking brain is that it works faster. But even with such advanced and compelling bodies of work in the field of behavioural economics, and with many Nobel Prizes within this very field of study, mainstream economists have made few concessions to account for biological factors.

We believe it is time for a serious and frank discussion of how the underlying human imperatives of the reptilian brain factor into national prosperity, by which we mean into the level of mobility countries foster. It is our contention that 'the reptilian always wins' – it always finds a way of manifesting itself – and if our national culture and institutions don't chime with our biology, mayhem ensues. Think about Kennedy and Marilyn, Clinton and Monica, Hollande and all of his mistresses. They all acted on their primal, reptilian instincts. Cultures where women are likely to die during childbirth naturally adopt polyandry, where one woman has several husbands. Why? Because in order to survive in that culture and in those conditions, and to be able to successfully reproduce, women rely on men for protection and nourishment. Thus polyandry becomes part of the survival kit.

Survival kits are inherited at birth and transmitted from one generation to another. Yet some cultures are frozen in time, and have stopped adapting. These cultures are at risk of extinction. This is what happened to the Greek and Egyptian civilizations: they relied on what had made them successful, but in failing to adapt, these very tools became the key contributors to their demise.

If our biological drives are thwarted and suppressed, we fall into a host of behaviours that are anti-mobility, that not only hold us back but others around us. So, if we are told we cannot fall in love with the girl down the street because she is Shiite and we are Sunni, our natural passion is suppressed and we are likely to act out through anger, indolence or perhaps through violence. Our mobility, and that of our whole culture, is hampered. If we are told that we can have only one child, and that it should be a boy, we may be driven to commit the ultimate anti-evolutionary act – infanticide. Our mobility, and that of our whole culture, is jeopardized. If we try to hail a cab and it drives past us because of the colour of our skin – and that's just one of the daily indignities of the racism we experience – we may become angry and shun mainstream culture, not believing in our potential for advancement. On the other hand, if our society supports us and urges us to find love however we choose to define it; if it helps us feel confident, cared for and respected; if it not only allows us to express ourselves but encourages us to do so; if it allows us to feel safe and to trust that we will have the medical care we need, it satisfies our basic biological drives, and that propels mobility, not only at the individual level, but for the society as a whole.

The movement of populations could also be seen as 'communicating vessels'. People from a culture which is moving down emigrate to one that is moving up. For instance, the Irish during the Great Famine moved en masse to America. Africans are now moving to Italy and Europe. Pakistanis move to Britain. In the past there were the barbarians looking for lands to conquer who attacked Rome, or the Ottomans who arrived at the gates of Vienna. The Arabs who were stopped at Poitiers by Charles Martel. The Crusaders who occupied Jerusalem, the British who occupied India, the Japanese who occupied Korea and China. This is not only a battle between soldiers and armies, it is a battle between cultures. So, for example, Korean culture was able to preserve its identity even as the Japanese dominated Korea for thirty-five years.

The Barbarian invaders, including Attila the Hun, who arrived at the gates of Rome contributed to the demise of the Roman Empire. Napoleon invading Venice ended the independence of that city state and its centuries of success. Hitler built an intricate network of highways in Germany so that he could quickly move troops and equipment in order to invade neighbouring countries efficiently. Pilgrim immigration to the New World changed the future of the American Indians. But if some 'American' cultures move up, some move down: Apaches, the Cheyenne, the Comanche and the Iroquois. Movement is always what change on hinges: it is the opposite of immobility.

The key lies in the relationship between culture and biology. Some cultures, at a certain time in history, were in harmony with biology. France in the eighteenth century and Moorish Spain in the twelfth century are two good examples. The danger arises when cultures crystallize temporary solutions or harden cultural norms that subsequently become obsolete. What was once an excellent solution to biological needs becomes a rule that destroys your chances of moving up. For instance, when more than 50 per cent of the world's population comprises women, forbidding women to receive a formal education, to work and to contribute to economic growth is utterly counterproductive.

Which brings us to the role of culture. The argument that culture plays an important, or even decisive, role in national prosperity has a long lineage, most often traced back to the pioneer of sociology Max Weber, who at the beginning of the twentieth century coined the term 'the Protestant work ethic' and credited it as a major factor in the success of the west.[7] Historian Niall Ferguson recently harked back to Weber in naming the work ethic as one of the six key factors he identifies as explaining the rise and predominance of the western nations in his bestselling *Civilization*.[8] David Landes, Jared Diamond, Jeffrey Sachs and many others have also credited cultural factors as playing an important role in development. But just as Weber's stress on the work ethic raised hackles for being a slur on the work ethic of the whole non-western world, attempts in general to explain how culture affects the development and the quality of life of countries have been accused of being biased by imperialism and racism. And it's true that too much assessment of national, regional and ethnic cultures has been tainted, whether consciously and intentionally so or not. Some scholars and authors who have extolled the virtues of liberal democracy and western capitalism have been rightfully accused of downplaying the problems in those systems and portraying other cultures with simplistic caricatures, such as that of the Latin '*mañana*' culture. Some have glorified a free market that is not nearly as free as they have described it as being, and have largely ignored many festering social problems and the rise of inequality in the western developed nations. On the other hand, some critics of the 'west is best' school of thought have been much too quick to underplay festering problems in the developing world, such as rampant corruption and outdated beliefs that run counter to social progress. Still others have flat-out denied the role of culture as summed up by James Robinson and Daron Acemoğlu in *Why Nations Fail*.

For us, there is no question that the nature of a country's culture plays a vital role in its ability to foster mobility. But what precisely do we mean by culture? Culture is how different groups of people systematically process the same information in their own way, leading to differences between groups in behaviour, traditional rituals, practices, attitudes and beliefs. We contend that these practices, attitudes and beliefs are either more or less conducive to mobility. Restricting people from marrying across racial and ethnic lines hinders mobility; the belief that God has called for women to be kept hidden in the home hinders mobility; the belief that certain races are by nature less intelligent or industrious than others hinders mobility; the practice of requiring business owners to pay bribes in order to obtain operating licences hinders mobility. The same goes for the practice of selected abortions on the basis of gender, or the belief that pleasure is a sin. By contrast, the belief that all people should have equal opportunity for empowerment and growth enhances mobility; as does the belief that education is a fundamental human right; the belief that sex is inherent to human nature; and the practice of promoting creativity. There can be no denying that some national cultures simply are more conducive to mobility than others. But there is no one culture that stands as a pure paragon of the 'good'.

The concept of the cultural unconscious plays an important role in our analysis. Cultures imprint individuals – defining and giving meaning to certain notions and experiences – and these cultural imprints are discoverable. We conducted an extensive analysis of how the culture codes of the world's nations either contribute to mobility or hold it back. We selected crucial aspects of the cultural unconscious that we argue affect mobility, such as the codes about freedom of choice, innovative spirit, gender equality, security, and the drive to succeed.

Each culture has words – codes – that define it.[9] Once we discover this cultural code, we suddenly understand that, through the collective cultural unconscious, one inherits a cultural personality, or cultural attitude towards life, and a certain expectation of the future. This is a complete reference system which has been imposed and imprinted at an early age and reinforced time and again throughout your lifetime. It is very interesting to see that, for example, German opera is very different from Italian opera. Love is different in Wagner operas to what it is in Puccini operas. The key verb for French culture is 'to think': 'I think, therefore I am.' In France you learn how to criticize other people's thinking in school with the *critique de texte* (literary critique). We found an incredible list of magazines and reviews dedicated to the French national sport of thinking: *Les Idées* ('Ideas'), *La Pensée* ('Thought') and *L'Esprit* ('Spirit') are all literary reviews. And there's Rodin's *The Thinker*; de Gaulle's influential phrase '*Une certaine idée de la France*' ('A certain idea of France'); the modern French hero, the philosopher Bernard-Henri Lévy; and TV shows about thinkers – *Apostrophes* with Bernard Pivot, for example, was popular for years.

By contrast, in the US, where the key verb is 'to do', action plans and 'to do' lists are very popular. In all the cultures we studied, we were looking for the reptilian dimension behind the collective cultural unconscious. We came to realize that the way a culture deals with sex, survival, security and success (what we call the Four S's) predetermines your potential to move up. This was a completely new way to look at mobility.

We know that life is movement, and that every culture that restricts movement, by extension restricts life. The Berlin Wall, anti-competition laws and immigration laws are a few examples. At the same time, cultures and nations need to protect themselves and preserve their identity. They don't want to open their borders to, for instance, the Ebola virus. You don't want your children running wild and never going to school, but on the other hand you don't want to restrict women to their homes and forbid girls to go to school: this does not help a culture to move up. So we explored basic human nature and how the reptilian always wins. How some cultures help the reptilian and others restrain it, and how the reptilian dimension can explain this. The relationship between biology and culture is the core of this book.

It's important to highlight that we are not arguing in favour of any one bio-logical nature or against any one cultural system. We are writing this book with the hope of bringing about positive change. Changes often happen slowly, and the first step is to become aware of the current cultural situation you are immersed in. When Turkey fought for independence, Mustafa Kemal Atatürk, the Turkish leader, began a process of modernizing the country, changing what needed to be changed. He ended the caliphate and separated religion from the government. He modernized education and invited foreign experts to teach in Turkish universities. He decentralized economic activity. Under his reforms, Turkish women gained political rights before those in many other countries. Mustafa Kemal borrowed successful structures from western countries and adapted them to the different realities of the Turkish people (who were mostly Muslim rather than Christian, for example).

Take the tradition of the dowry. In many cultures, when a couple gets married social rules require the bride's family to offer a substantial amount of cash or other gifts to the husband's family, the traditional purpose being to consolidate the bond between the two families. But today some families find it such a burden to be expected to pay a dowry that this has led to female infanticide. The more mobile cultures alter or abandon traditions found to have such pernicious effects.

In Chapter 12, having devised our analysis of the mobility of nations' cultures, we will rank countries by how well they enable their people to satisfy, or channel, their biological drives. The better they do so, the more they are fostering mobility. These drives are not only for survival, but for pleasure, love, self-expression, respect, intellectual stimulation and growth, a sense of purpose, and personal expression.

Rather than subcribing to Maslow's hierarchy of needs, we devised our own breakdown of the biological imperatives into our Four S's: survival, sex, security and success. Survival is about the degree to which the country supports the health, education and overall well-being of its people; sex is about the degree of gender equality, and how pleasure is conceived differently across cultures; security is about how well the country offers protection from physical, economic and environmental harm; and success is about how well its institutions foster economic competitiveness, efficiency and innovation.

While we have no quibble with Maslow's argument that survival and reproduction are the most basic needs, we think it's important to understand that those he identified as being higher up the pyramid are also fundamentally biological. Evolutionary psychology has convincingly argued, for example, that our drive for success is closely related to the many benefits success brings in assuring our access to the basic requirements of survival, as well as to our ability to bond with our preferred mates. The notion of the pyramid seemed to suggest that we can rise above our basic biology. We want to stress that our basic biology is always driving us, even in the work we choose to do, our appreciation or creation of art, our pursuit of learning, and our offerings to charity.

Our argument is that the better a country is at fostering the satisfaction of these biological needs, the more conducive to mobility it is. We decided to devise quantitative measures of how well countries do this, and for that we conducted a multi-year in-depth assessment of the wide range of existing measures of national prosperity and human development created by scholars and by organizations, such as *The Economist* Intelligence Unit, the World Economic Forum, the Heritage Foundation, the *Wall Street Journal* and the United Nations.

When we came up with this index, our intention was to allow individuals to benchmark themselves. Our results are of course not politically correct, and may be unsettling. Now that you are aware of how your culture or national identity pre-programs you to succeed or to fail, you are able to discern what you can do about this if you are inclined to move up. You might just want to vote with your feet. The excuse used by the French, for example, that their problems stem from globalization and the economic crisis doesn't explain why the Germans aren't experiencing the same problems. Germany is doing well economically and has a much lower unemployment rate than France.

So you might choose, like many young French entrepreneurs, to move to Berlin or London, where there is a more business-friendly environment and more chances to move up. This is not just the traditional brain drain that we have observed for centuries, it is a new awareness: you can pick and choose where in the world you personally can thrive.

Some countries might make it difficult for you to leave, for instance, North Korea, or Cuba before Raúl Castro loosened the travel restrictions. Other countries, for example Switzerland, will put up barriers to entry. If people want to leave their country, this is a clear sign that they feel as though they cannot move up in their culture, and believe that they may have a greater chance of doing so somewhere else.

Herbert Spencer wrote about the survival of the fittest in his book *Principles of Biology*, referring to Darwin's idea of natural selection.[10] In our book, we discuss the survival of the 'up' culture, a culture where people can prosper. Some people may call this upward mobility. We will see that some cultures that offer equal opportunity, predictability and flexibility have greater chances of moving up. We hope that this book will stimulate people around the world to move, and to move up.

A NEW PARADIGM IS BORN

1 *Back to Basics*

Man can do what he wills,
but he cannot will what he wills.

Arthur Schopenhauer[1]

Let's talk about love.

Richard and Megan met briefly during orientation on their student exchange programme in Madrid. Though attracted to one another, they had just arrived in Madrid and were more excited about the prospect of the many, potentially more attractive, people to choose from. They didn't cross paths again until the end of the semester when all the exchange students fled to Barcelona in search of fun before returning home.

Megan and Richard sat next to each other on the train and glanced at each other, impressed but not quite sure. For the three-hour train ride they engaged in subtle flirting. Megan's bare leg slightly rubbed against Richard's, arousing him, and the scent of his sweat sent hormones darting through Megan's body. Their hearts raced, their body temperatures rose, and Megan had a slight pinkish glow in her cheeks. As soon as the train stopped, they both stood up and had a better look at each other. They really liked what they saw. Richard helped Megan get her luggage down and made small talk on their way to the hostel where everyone was staying.

The rest of their stay in Barcelona was magical. They spent every waking hour together, and by the third day they got a private room at the hostel. Their chemistry was undeniable and their attraction magnetic. The charm and elegance of the city was the perfect backdrop to long strolls through the night and romantic stargazing. Their last days were bittersweet. They were sad they would have to part, but made the best of their last moments together.

On their last day, they rose with the sun and Megan packed for her flight back to San Francisco. Richard took her to the airport and they promised to keep in touch. They shed a tear, hugged for eternity, kissed and parted. Months and years went by. Each of them lived their own life, but, although they had other relationships, they would never forget that perfect encounter.

Three years later Megan organized a trip to Europe with her friend Kate. Their first stop was Paris, where they had planned to spend a week catching up with friends. But then Megan decided to drop in on Richard in London. When she arrived at the station, she quickly spotted him walking towards her, hesitant and unsure. He drew close. 'Megan?' Like a wonderfully orchestrated game of destiny they were once again in the same place at the same time. They hadn't been sure if they would recognize or even like each other any more, but it took just a few minutes for them to start experiencing that same sense of excitement and warmth they had shared in Barcelona.

Megan ended up staying a week longer than planned. They were passionate with each other and found that in every way they just felt at home: safe. At the end of Megan's stay, they decided to make plans to move to the same city. Richard was finishing his course and preparing to move back to the US – he was happy to give San Francisco a go. Megan caught her train back to Paris with a huge smile on her face: she had found her perfect match and this was only the beginning. She had fallen in love with him and felt they could truly build an amazing life together. She would have stayed longer but she had promised Kate that she would attend her Parisian birthday celebration, so she hurried back.

At this point in the story everything seemed promising for the passionate couple, their exciting future unfolding in Megan's mind as she made her way to the party. She could not stop thinking about Richard.

That night, Megan met Bradley. He was handsome and strong, and exuded confidence. An up-and-coming Australian professional rugby player, Bradley was celebrating his team's victory: he had scored twice in a match with France. He was very tall, had sun-kissed skin, golden curls and probably the most beautiful smile she had ever seen. Bradley was attracted to Megan as soon as he saw her, and, although she held back at the beginning, after a few glasses of wine and twirls on the dance floor she found it hard to resist his charm. He took her to his hotel suite where he seduced her. She would never see him again.

The next day, Megan received a few texts from Richard and a long and heartfelt email, telling her that he had never been ready to commit to a relationship before, but that he was ready now. Megan thought she needed to be completely honest for the sake of the relationship, and because she was truly confused. So she called Richard and confessed everything. Richard reacted with sadness and rage, ultimately hanging up on her. Megan didn't hear from him for two weeks, so she decided to call him. Another woman answered and passed the phone to Richard. He was brief and cold. He admitted that he didn't have it in him to give Megan another chance. So they never saw one another again.

Why did Megan cheat on Richard despite the strength of her feelings for him? Why was she so easily seduced if she believed she had found the love of her life? Her future was laid out before her, but her sex drive was more powerful.[2] In going off with Bradley she was obeying her natural instincts, perhaps for an ego boost; the hottest man in the room was attracted specifically to her. She probably did it because she was attracted to his scent, facial symmetry and strong build.[3] She probably just wanted to feel a sense of adventure, something different. Or maybe she didn't love Richard enough. She didn't lose the love of her life because she cheated, though; she lost him because she told the truth.

Any one of us could have been Richard or Megan in this story, and many of us have probably been in either or both positions. Monogamy is a social construct, and as humans, who are nothing more than mere specimens of life, we love to cheat, and not only on our romantic partners. Whether it's a college test, friendly soccer match or taking credit for work we didn't do, more often than not we are going to cheat a little here and there.

"You slept with her, didn't you?"

But we are surprisingly irrational when it comes to our cheating and scheming. When the reward from cheating is big, we feel more guilty about it and more reluctant to think of ourselves as cheats. That is why the bigger the reward, the less likely we are to cheat. We prefer cheating many times in small ways (when we feel less guilty) to cheating occasionally in a big way (when the guilt is overwhelming).[4] This is where those cultural values come into play: guilt and conscience. It perfectly explains why Megan would find it wise to confess to Richard, despite the consequences.

So many films, novels, poems and songs have told stories of falling in and out of love.[5] However, what we forget to look at are the reasons people do the things they do. We are all biological beings with innate tendencies, needs and desires, but what happens when our human instincts contradict our best interests, when the reptilian brain is up against the cortex? Megan's desire for the alpha male at the party led her to lose probably the love of her life, someone with whom she could have built a family and lived a happy and successful life. What happens when your own nature betrays you and leads you to behaviours that are actually counterproductive to moving up? How can we understand these dichotomies between our reptilian self and our cortex?

The battle between our bio-logical drives and reason, between destiny and choice, are battles all humans and all cultures face. Why 'bio-logical'? Because we need to be logical about our biology. This book is an attempt to unveil why this battle is no longer necessary, and really never was. If we want to move up, our best bet is understanding, accepting and even celebrating the beast within, our reptilian self, because, whether we like it or not, the reptilian always wins.

*

Now let's talk about China.

With its rich history, varied cultures and traditions, twentieth-century China was a particularly delicate and interesting place. A future leader was born into the turmoil of the turn of that century. By the time Deng Xiaoping was in his teens, the Qing Dynasty had fallen, and the political future was uncertain. He spent his youth studying in France and Russia, greatly influenced by his Communist classmates, and prepared to enter China's political sphere.

By 1949, when Mao proclaimed the People's Republic of China, the old traditional culture of China had broken down after many years of war and famine. Everything that was once held sacred was now forbidden. In the new culture private property was banned; individualism and freedom of expression were shunned; western economies were considered poisonous; and to be rich and successful was seen as abusive. This new hybrid of Chinese Communist culture shunned anything traditional, whether Chinese or western.

As he made his way up the political ranks, Deng managed to reinforce his loyalty to Mao and at the same time challenge the system Mao had established. He took over as leader after Mao's death in 1980.

Imagine growing up during China's Cultural Revolution, closed to the rest of the world and forced into a Communist market, and one day waking up to a China that was open. But Deng's success was not based on erasing China's cultural heritage – which would barely have been possible. What he did instead was adapt traditional Chinese culture to modern times. He kept the Communist rhetoric and group-oriented style that was deeply rooted, but opened the doors to the free market economy in order for China to compete on a global scale. 'One country, two systems,' he once said, referring to

China's ability to remain Communist while adapting to a market economy.

His reforms, or the Four Modernizations as they are known, ranged from agriculture to innovation, trade to industry, and even the military. Once he was able to open China to the world, the country grew exponentially, and continues to do so. Farmers were able to capitalize the land and market products, reducing the damage done by China's Great Leap Forward and the Cultural Revolution. China also opened up to new technologies and innovations from the western world after decades of being closed off from international advances in science. The barriers to trade were dropped and China was able to import technologies and export products, inviting greater investment into the country.[6]

Since Deng's time, China has become one of the strongest players on the international economic playing field. Compared to Mao's closed society, Deng opened China to the world. How was China able to grow so quickly, economically and socially? How were the Chinese able to adopt competitiveness while keeping their cultural heritage? How were they able to transition from strong Communist ideals to those of the free market and meritocracy in such a short period of time? How were they able to move up despite such a strong and difficult past?

There is always the question of what traits help a culture to foster mobility. In 1947, when the United States was attempting to formulate a universal declaration of human rights, the American Anthropological Association stepped forward and said, 'It can't be done; this would be to impose one provincial notion of human rights.'

How can we persuade all of the people who are committed to self-destructive and harmful things in the name of 'culture' to change their beliefs, to have different goals in life, and to lead better lives?

How can we understand philosophers and scientists talking about the 'contextual legitimacy' of female genital mutilation, or any other barbaric practice that we know causes degradation, misery and social paralysis?

We are saying there are cultures that encourage people to flourish, and others that do not. In China, under Deng Xiaoping, the lesson was cultural: changing a country's culture code changed everything else – it had a domino effect on nearly all aspects of life. With that concept in mind, how can we think about cultural mobility in the context of science?

Biology sees no distinction between good and evil, it only knows what works and what doesn't, what helps us survive and what doesn't. But we aren't just concerned about whether our culture helps us survive, we want to know if it helps us to grow. Our world-view is not universal; who is to say one culture is better than another? What suits a Manchurian cattle herder may not suit a Swedish schoolteacher. What we do know is that it is possible to create a benchmarking system for comparing different aspects of cultures to see which aspects help us move up and which just bring us down.

Our explanation of the universe – our sets of values, principles and descriptions that explain the purpose of life and the meaning of our actions – varies between cultures. For example, the United States takes Thomas Jefferson's Declaration of Independence as completely factual and self-evident. But these values may not be self-evident for people in Nepal or Haiti. Each culture has its own universe or world vision. The world conceived by Confucius or Buddha is not the world conceived by Voltaire or Descartes.

We also believe in the relativity of culture, and this book

will explore why cultures develop the way they do, why some succeed and why others fail and disappear. So the core of this book is the relationship between biology and culture. We are not interested in making a choice between the two, between nature and nurture. Rather, we are interested in their interaction.

We've found that a culture that goes against our biology does not last very long. To put it bluntly, a culture that tells you to pick up a gun in the morning and kill is not going to survive. Cultures that get stuck in old situations, or old solutions, that do not fit into their new environment or adapt to new problems, will decay and disappear.

Where does this leave us with mobility? Well, we need a culture that works for our biology, and a biology that works for our culture. Let's put it in a slightly different way. Sometimes there are conflicts between biology and culture. For example, there is still a natural selfishness in the human brain, and cultures need to learn to overcome this natural impulse to some degree. Cultures need to make us understand that we need to cooperate not only for the sake of the group but for the sake of every individual in the long term. In the real world, this means that, in order for a culture to move up, you have to keep your promises, you have to pay your taxes, you have to eat healthier, you have to wait your turn in the queue, and maybe you even have to send your children to war and risk their lives.[7]

The same thing happens with our sex drive and our cravings. Perhaps you will have to limit yourself by not having sex with an intern if you are the president of your country and you are married. It might go against your biology, but these are the sorts of things that cultures promote in order for their people to repress their natural selfish impulses and do things that help their societies to move up. This benefits everyone in the long run. (Cultures do this, and of course some laws and moral codes too.)

This is not to say that all group-oriented behaviour is cultural. There are some biological traits that evolved to enable people to overcome impulses that are likely to destroy the group and to do what's best for the group instead, including self-control, self-regulation and fairness instincts. Later cultural developments served to replace aggression (a biological instinct) with morals and laws as the primary means to solve conflict.

If biology has to be bio-logical to foster mobility, culture also has to work with our biological make-up. This is the key for mobility: one supports the other.

We are not attempting to make a value judgement about Megan and Richard. It's not whether they were right or wrong in cheating or in making a promise they could not keep. They just needed to acknowledge their impulses and not simply ignore them. Ignoring them doesn't make them go away. If you want to keep a promise to yourself and to another person, you need to understand the way the reptilian brain works and find a way to express it rather than repress it. The lesson with Megan and Richard was bio-logical, being logical about our biology by both listening to and understanding it. Being aware of our instincts is the key to developing a culture that can harmonize rather than clash with biology.

THE HUMAN ANIMAL

2 *The Reptilian Always Wins*

I have been and still am a seeker, but I have ceased to
question stars and books; I have begun to listen to
the teaching my blood whispers to me.

Hermann Hesse, *Demian*[1]

In one of the grandest and most beautiful homes in America, there lives a handsome, bright and charming man with a lot of power. At one point, knowing that he may lose all of this power and his reputation, he begins an affair with a young woman. So goes the story of Bill Clinton and his intern, Monica Lewinsky.

How can you let a situation like that get the better of you when you are presiding over the most powerful country in the world? At this level of power, it's impossible to have any secrets, and yet he let his hormones dictate to him. The consequences were an impeachment and one of the biggest public scandals any American president has ever faced. Clinton is a brilliant man but there's no stopping the reptilian brain. Maybe it was his brilliance that led him to cheat. Talented people are usually more creative, and creative people are more likely to cheat, since they are better at deceiving themselves into thinking that what they are doing isn't wrong. That's the reptilian, and it always wins.[2]

"I HOPE I'LL BE REMEMBERED AS A SUCCESSFUL TWO-TIMER... ER, TERMER."

The Three Brains

Let's start by getting to know ourselves – and our brains – a bit. The brain is obviously an extremely complex structure: neuroscientists are the first to admit we only know the tip of the iceberg. An illustration of how the brain works could be simplified into a map, which is a high-level model of how the brain is wired.

If I give you a map of France it is not the territory of France, but if you want to go from Marseille to Lyon, you'd better have a map. So if you want to understand the brain and how people behave and function, it's better to have a map. The American neuroscientist Paul MacLean developed an ingenious way of mapping out the brain in a simplistic and comprehensive way in order to understand why we do what we do. He called this the 'triune brain' model, since it's made up of the cortex, the limbic and the reptilian brains.[3]

All human brains have three basic structures. The cortex is where complex mental processes occur, including language acquisition, planning, abstraction and perception. Some even say that this is the truly human part of the brain. The limbic brain supports functions that include behaviour, emotion, memory and motivation. The reptilian brain is responsible for instinctive behaviours shared across species which include

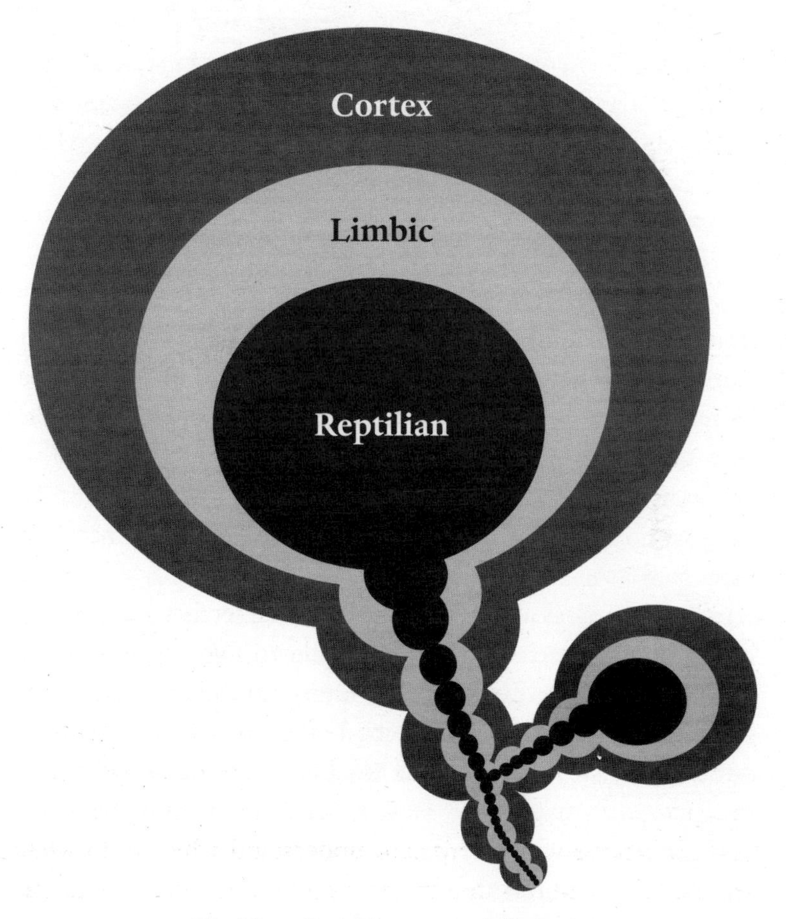

The Three Basic Structures of the Brain

aggression, dominance and territoriality, as well as many of our body's vital functions such as breathing, body temperature and balance.

The undisputed champion of the three 'brains' is the reptilian brain. The name comes from the region's similarity to the brains of reptiles, which are believed to be relatively unchanged from the brains their predecessors had 200 million years ago. Our reptilian brain programs us for two major things: survival and reproduction. These are, of course, our most fundamental instincts: if we could not survive and reproduce, our species would be extinguished. It is the part of your brain that allows you to dodge a ball aimed at your head, get an erection when aroused by an attractive woman, and gulp down water when thirsty.[4]

The reptilian brain can overpower our other two brains. Physical attraction, for instance, has a strong reptilian dimension. At the reptilian level, you are physically attracted to someone whose genes provide the best chances of survival for your children, so you will look for things like physical health, material resources, kindness and intelligence.[5]

The way in which the reptilian brain interacts with our limbic brain and cortex determines how our basic instincts are expressed in our behaviour. But when functioning alone, the reptilian brain has no sense of right or wrong, feeling or thinking: it simply receives stimuli from our senses and responds. Our minds have evolved, just like our other organs, in order to ensure our reproductive success. This means that sometimes our conscience, or doing what is right, is not necessarily conducive to our survival – to moving up. Clinton knew that if he told the truth about his affair he risked his entire career. Lies and deceit are crucial strategies for survival: they cover up the reptilian tracks that our cultures disagree with. This was

Clinton's cortex at work, consciously making the decision to lie in order to avoid the consequences of his actions.[6]

If we look at simple things we take for granted, we can easily link them back to the reptilian. Soccer is completely reptilian. It reminds us of the primitive hunt that involves chasing a target, working as a team in order to win, developing a strategy and honing our skills. It's linked strongly to our emotions, with the euphoria of winning, the depression of losing, and the sense of belonging to a team.[7]

The reptilian brain is what motivated the surviving Uruguayan rugby players to eat the remains of their fellow team-mates when no other food source was available. It's what inspires Silvio Berlusconi to engage in extramarital affairs, even with under-age women. Mike Tyson's reptilian brain pushed him to bite Evander Holyfield's ear off, which lost him the boxing match (and his credibility as a respectable athlete).

But, not only do we have three brains, our cultures also have three brains. Cultures have aspects that are reptilian, limbic and cortical, and they also deal with these three parts of us in different ways.

Diamond in the Rough: The Reptilian

When Megan cheated on Richard, her reptilian brain took charge. She didn't think it through: her body was craving Bradley. The reptilian won. We are born with a reptilian brain, it is not acquired, and every culture has it too. Regardless of differences between cultures, the reptilian part of each culture is universal. Reproducing our genes is reptilian, and therefore universal. *How* we have children, create a family and raise them is what differs from culture to culture. The reptilian may be universal, but the ways in which each culture deals with it are unique.

Men and women belong to different genders which are truly disparate. Men are programmed for quantity. Why is this? Cro-Magnon man didn't live very long. He had a limited amount of time to reproduce his genes, which is why men have testes that produce 400 billion sperm over the course of their lifetime.[8] Not only do men's sperm have to compete for the prized egg, men too have to compete for sexual access to the prized woman, so naturally more is better because it means you have a greater chance of reproducing your genes. This is completely reptilian.

But men's lifespan today is much longer than it was over 40,000 years ago. So why do men still produce so much sperm and seek so much sex even with so much more time to reproduce their genes? It is because the reptilian brain evolves very slowly, taking millions of years to change its programming. In contrast, cultures can evolve in a few generations. There is a huge disconnect here! On the one hand it is undeniable that we have a brain that is still programmed to lust, and to satisfy its lusts; and on the other hand we have constantly changing cultures that tell us how to deal with this reptilian need.

This gap between the evolution of the reptilian and the evolution of our cultures gives rise to all kinds of miscommunication. But we've found the solution: those cultures that are best suited to survive are those that accept the reptilian dimension rather than repress it. You are entitled to be aroused by a person of the opposite sex but you also have the choice to control your desire.

On the other side of the coin, women are programmed for quality. Whilst men can literally have sex with an attractive woman one night and a not so attractive woman the next, women are much more choosy. When a group of women go out with the intention of having some fun at a bar or nightclub, most of them would rather go home alone than sleep with a man below their standards. This is because women are the gatekeepers, protecting their uterus from unwanted suitors. Pregnancy, childbirth and child-rearing are huge investments on her part, but not so much for a man, so it is in her best interest to choose the best possible option available. She prefers quality over quantity.[9]

But, contrary to the belief passed on to us from centuries of oppression, female sexuality is no less potent than male sexuality. It is culture that has denied the reptilian side of women through the stigmatization of promiscuity, through female genital mutilation, and so forth. In fact, only a few decades ago the female orgasm was still perceived as a myth, or worse, doctors actively searched for a cure for 'female hysteria'. Many societies around the world can be characterized as being

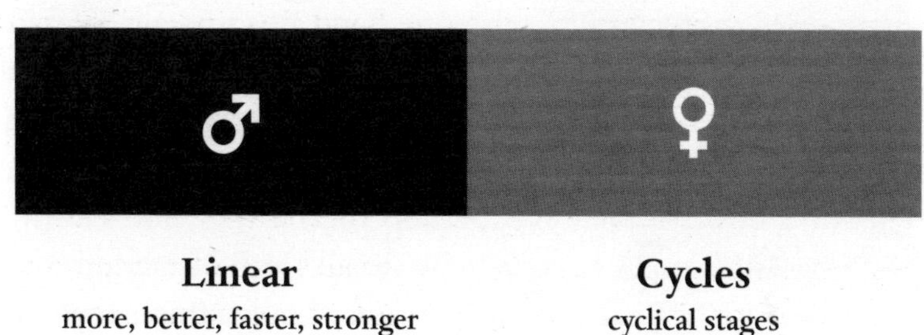

♂	♀
Quantity 400 billion sperm	**Quality** 1.5 million eggs (but only 400 mature)
Outside knocks on the door	**Inside** door keeper
Now immediate, in the moment	**Delay** patience, timing

Men v. Women

'macho', where men are praised for having multiple lovers, while women are socially prohibited from having the same privilege, even if there is no logical reason for two people who are attracted to each other not to enjoy each other sexually. These rules and disparities, taboos of the sacred and the profane, are our culture's limbic side.

This brings us to yet another reptilian gap between men and women, to do with orgasm. Men get excited and reach orgasm very quickly, following a linear curve until they climax. After sex, a man's biochemistry changes, releasing prolactin which makes him feel sleepy.[10] Women on the other hand have fluctuations of pleasure during sex: one second they are excited, the next they are bored. But, once they reach their peak, they can go on for a long time and even have multiple orgasms.[11]

The disconnect between a man's climax and a woman's climax is the reptilian gap, where women need to get excited more quickly and men need to go more slowly. This reptilian gap in sexual behaviour and mate-seeking creates a serious conflict of interest between men and women. With such a dilemma, how will they ever get together?

These differences between the female reptilian brain and the male reptilian brain may be responsible for the many discrepancies evolutionary psychologists have found in how men and women approach the world. Women are more connected with their emotions and interested in people whereas men are more drawn to objects; women rely more on words, men are more likely to visualize a problem; women focus on reciprocity in relationships, men pay attention to who is dominant; women have a better understanding of the space inside, and men are more into exterior space. Some people even say that men may water the garden and mow the lawn, but women plant the flowers.

The views expressed here may be perceived as sexist, and not politically correct, but it is clear that cultures which do not understand and reflect the unconscious biological structure (including those differences between men and women) do not help their members to move up. There is a new movement in the US who's slogan is 'let men be men, and women be women'. It is interesting that some people in America feel the need to reaffirm the differences between men and women!

Since both sexes are so inherently different, it is important for them to understand each other and their sexuality. In Indian culture, for instance in the *Kama Sutra*, men are taught how to take their time and pleasure a woman until she climaxes. Cultures need to be more reptilian rather than moral about such things.[12]

When we analyse the reptilian, we are looking at it through a scientific lens. It is true that beliefs and superstitions about the unknown are very powerful, but science can have profound implications on how we perceive our world. For centuries, cultures interpreted the female orgasm in different ways, from denying its existence to attributing it to female hysteria, but through recent scientific discoveries we are able to unveil its true nature. These types of scientific discoveries can help us better understand each other.[13]

Unsurprisingly, men and women develop very differently. A man's growth throughout his lifetime is very simple and linear. It increases, and then at a certain point starts to decrease. For women, it is much more nuanced.

Women's physical evolution is cyclical. First, they are babies, then they are little girls, then they go through puberty, then they become sexually active, then they become pregnant, then become a mother, then enter menopause, and then a whole new identity begins. They self-identify differently at each stage.

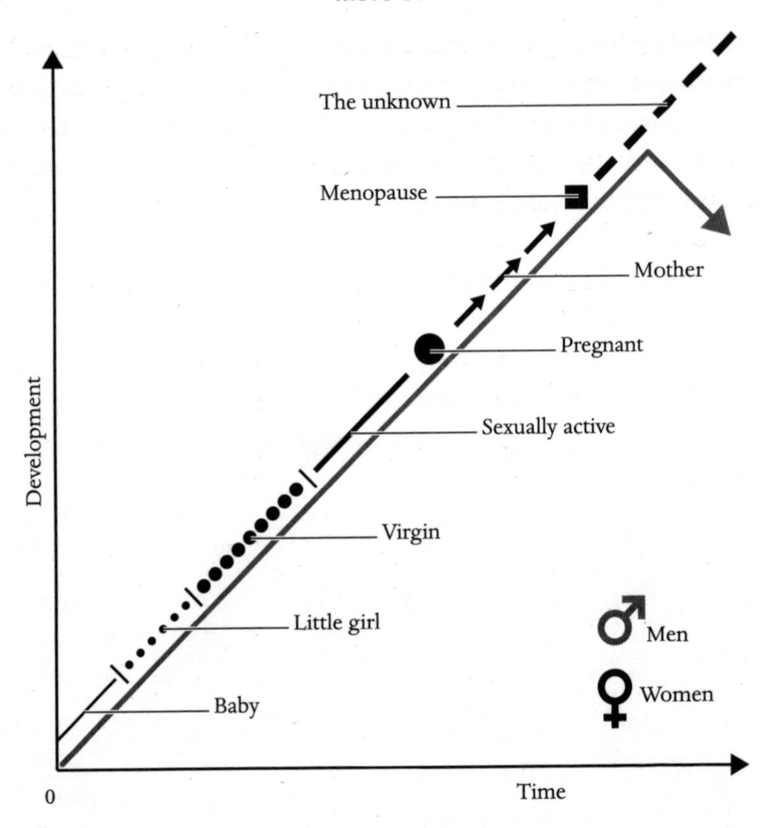

Human Development

Men are a lot simpler – they just want to have sex. Men do not go through so many biological transformations as women. In many ways, men are simpler, biologically speaking. Men start out as babies of course; they then become boys, then adolescents, and finally become men by going through a series of rituals in order to prove their identity. Drinking, fighting, marriage and then fatherhood, after which they do not have nearly as much sex as they wish they could.

The reptilian gap here is that a man marries a woman

thinking that she will never change, and she always changes. A woman marries a man thinking that she can change him, and men never change. Natalie Wood once said, 'The only time a woman really succeeds in changing a man is when he is a baby,' and she wasn't wrong.[14]

Learning How to Polish: The Limbic

> There are as many ways to live in this world as there are people in this world, and each one deserves a closer look.
>
> Ole Golly in the film *Harriet the Spy* (1996)

After the night that Megan spent with Bradley, she somehow felt she had to confess. She felt Richard deserved the truth. It certainly crossed her mind not to tell him; after all, she wasn't planning on seeing Bradley again. Also, Richard would have never found out. Megan had an emotional battle, and she finally decided, from what she had been taught in her culture, that honesty was the best route. This was the limbic working. The limbic part of the brain is acquired; it's what our cultures teach us. It's because of the limbic part of our cultures that every way of life is possible.

In Tibetan and Eskimo cultures, if a young girl comes home and tells her mother, 'Guess what, Mum! I'm in love!' the first thing her mother will ask her is, 'With how many men?' Being in love with only one man is seen as naive; if she's in love with three men it's fantastic. This is because, in these cultures, polyandry is accepted. Why? Because the culture created a solution that suited its environment best in order to survive. In an environment where there are fewer women than men, and most of the time women die during childbirth, polyandry is a good option.[15]

In some African cultures, a guest is encouraged to sleep with the host's wife in order to relieve any sexual tension or threat of dominance right off the bat. In some western cultures, it is common for teenagers to experiment sexually with other teenagers while their parents are absent, at a party or even in a parked car. In some Middle Eastern cultures, however, it is common for a teenage boy to be initiated into sex by an older woman, and a teenage girl by an older man – more experienced partners who will teach them all about sex.

Anyone who has had the opportunity to travel to the extremities of the earth has probably noticed a key difference between cultures: toilets. In China, people use toilets that are essentially holes in the ground, and privacy isn't really an issue: the bathroom stall walls are often waist-height and people rarely close the door. In most Lebanese households you'll find a small bottle of water next to the toilet to use while cleaning your rear end. Westerners might find it odd, but Lebanese think westerners dirty for limiting their cleaning ritual to dry toilet paper.

In his book *A History of the World in 6 Glasses*, Tom Standage highlights how different beverages are closely tied to history and can be seen as a way to 'demonstrate the complex interplay of different civilizations and interconnectedness of world cultures'.[16] For example, in 4000 BCE beer was the beverage of choice. Its power was vast, as seen in pictograms from Mesopotamia. Those who built the Pyramids were paid in beer and bread, and it quickly became the drink of the working class, and to an extent still is.

Wine however has been associated with the social elite and religious activities since its origin. A region with vineyards was perceived as being economically powerful. During the Age of Exploration, spirits were the poison of choice: sailors on long voyages would make grog by mixing rum, water and lemon juice, improving the flavour and reducing their risk of scurvy or other illnesses.

Coffee, on the other hand, 'rose as an alternative to alcohol, and coffeehouses as alternatives to taverns – both of which are banned by Islam', spreading throughout the Arab world during the fifteenth century. In the eighteenth century, Europe welcomed coffee during the Age of Reason as a drink that sharpened the mind.

At one point, our world was defined by Coca Cola, the universal drink. What Coca Cola represents is so much more than just a beverage: it represents economic power and wealth. Regardless of your circumstances, a Coca Cola feels like a luxury, and provides a sensation of well-being. But each culture drinks a Coca Cola recipe specific to their area. The variety is incredibly vast: our 'universe' is not universal. So, in essence, cultures take reptilian-like behaviours, such as sex, and the limbic part of the culture adapts them to the local needs. They are solutions that cultures create at a given point in time, always keeping in mind our essential need to reproduce our genes.

Polishing the Diamond: The Cortex

Now let's talk about the cortex and culture. Our cortex is the rational and logical part of our brain that we acquire around the age of seven, and cultures have it too. It is the world of engineers and higher pleasures. Using the cortex, our cultures take our basic reptilian needs, accept them, then make them beautiful and civilized, and give them their proper place in our lives. From eating to sleeping, from having sex to going to the toilet, all are reptilian behaviours; and it's the cortex's job to turn these basic needs into an art, a higher pleasure.[17]

When the Portuguese colonizers arrived in Brazil, they

found the local women frolicking naked on the beach, happy, open and sexually available. Coming from Spain and Portugal, at the time very sexually repressive cultures, the colonizers were shocked but pleased: it was paradise for them. Today's Brazilian culture is the epitome of non-repression, of expression, of accepting and celebrating the reptilian inside us all. It is very feminine; it is a culture that celebrates beauty, sex, the human body, dancing, eating, pleasure and music. This is their cortex at work, dealing with the reptilian and turning it into a pleasurable art, which in turn makes the society non-aggressive.

Cultures around the world deal with personal hygiene in many different ways. Public baths in Ancient Greece were places of repose after exercising or thinking, a place where people could socialize and enjoy bathing together with warm water, aromatic oils and beautiful gardens. The Misogi in Japanese culture ritualized bathing as a process of purification, originally with a religious meaning but later becoming a leisure activity. Today, many cultures shun public nudity and make it a source of guilt. What public baths did was take away the guilt from our reptilian needs and celebrate them: that's healthy.

In America, you can eat for less than a dollar at McDonald's, and people eat a lot of poor-quality food at such low prices that most of the population is now obese. But imagine if sex cost less than a dollar and was as widely available as food: we'd probably have far fewer obese people. In America, food is safe sex. Unfortunately, America is not famous for its refined cuisine.

In 2006, the world was shocked by Norway's announcement that they were going to create a pension fund for future generations based on earnings from recent oil discoveries. This is long-term, cortex thinking. In a short-term world, natural resources are to be used now and benefit people today. It's like the saying 'finders keepers, losers weepers'. But Norway has a very cortex mindset: they perceive natural resources to be truly national, and therefore the benefits belong to all Norwegians, even future Norwegians.[18]

But there is another interesting dimension to the three brains, and it has to do with time, space and energy. Each are dealt with differently from culture to culture, and this depends on whether the reptilian, limbic or cortex dominates.

3 *Time, Space and Energy*

The Triangle Inside Our Minds: Time, Space and Energy

When we think about time, what do we think about? Are we thinking in minutes, hours, days, months or years? What defines the way we understand our experiences in terms of time?

And what about space? If the universe is in fact infinite, how are we supposed to define space? Or more difficult still: how can our brain have three different parts yet have no space between them?

Finally we have energy. Thanks to the law of conservation of energy we know that energy can neither be created nor destroyed, only converted from one form to another. So if energy has no end or beginning, how can we track the forces or the substances that keep energy alive?

We have been inspired by one of our greatest intellectual ancestors, Edward T. Hall, who spent most of his life studying concepts of time and space in various cultures.[1] This chapter explores how the reptilian, limbic and cortex interpret time, space and energy, and how this in turn affects the way cultures behave and help people move up.

Time

Time for the reptilian is now, always seeking immediate satisfaction any way possible. The reptilian speaks to you when your mouth waters as you stare at a plate of ribs, tempting you to dive in right now. It speaks to you when you are in the car on a long road trip and urging the driver to pull over before your bladder bursts. Patience isn't part of the reptilian.

In today's globalized world, borders can be transcended thanks to the Internet. People are more connected today than ever before. When only a century ago lovers might wait weeks for a letter to arrive from overseas, today a virtual conversation between opposite corners of the globe can take place within seconds. With social media like Twitter and Facebook, communication and information is instantaneous: it's reptilian time, in the now. Why is this? Because people love hearing and talking about themselves, and they especially love receiving positive feedback from others. It's instant gratification in real time.[2]

Have you ever wondered why Facebook has over one billion users? It's because Facebook is absolutely reptilian. It reinforces the reptilian tribe mentality, that feeling of belonging to a community and being valued within that space. Your friends and groups are like tribes, you follow people (and get followed), you become a fan, you belong to networks and you are always seeking to expand them. Rituals are created, rituals that govern the way you share information, establish and foster relationships, participate in groups and events, 'hunt' for mates, and so forth. And, like the tribe, it's about gaining territory, having the most friends, photo likes and comments on your wall.

But, like the tribe that hunts, there is always a risk, a possibility of being shamed or rejected. 'Should I respond now or later? Will I look desperate? If I share this song will he notice? What does my profile picture say about me?' These are reptilian questions, which we ask ourselves because ultimately we want to be accepted and acknowledged.

In American culture, time is very reptilian: focused on the short term. If we look back, the American settlers escaped their past, erased it, started afresh and made a new life for themselves. They burnt their bridges, emancipating themselves from the stranglehold of the English monarchy and Church, and seeking to start anew. Their debts, history, heritage, even their names were erased. For the first Americans, time was not about yesterday or tomorrow: their history was forgotten, their future a blank page, and time started now.

For the limbic, time is about delay. It's what tells you not to gorge on the bread basket at a restaurant when you know your boeuf bourguignon is on its way. Not quite able to see into the distant past or future, the limbic is into the short term. Unlike the reptilian, the limbic can delay pleasure, thinking that the future reward will be better. Agricultural societies perceive time in the same way, making the conscious decision to plant seeds now because they can expect them to yield crops in a few months.

The famous Stanford marshmallow experiment demonstrates the difference between reptilian and limbic time. In the study, children under the age of five were given a marshmallow in a controlled environment. The children were told that if they restrained themselves from eating the marshmallow, they would be given an additional marshmallow within fifteen minutes as a reward. There were two outcomes: children who were more reptilian ate the single marshmallow instantly, whereas children who were limbic delayed their pleasure in the expectation of an even greater reward. Follow-up studies conducted on the children years later revealed that those children who delayed gratification tended to score higher on their SATs and were described as being more competent by their parents. Understanding the reptilian is important, but the delay of the limbic also has its value in certain situations.[3]

For many Latin American countries, time is about delay and

is therefore limbic. Things that need to get done are not done in the moment, rather they are delayed, postponed for tomorrow, or '*mañana*'. 'Do you need to pick up that bag you left at my home? I'll drop it off *mañana* . . . I owe you money? I'll pay you back *mañana* . . . Do the kids want to go to the beach? Not today, I'm tired, better *mañana*.' Like a sleepy beach town, urgency doesn't exist in limbic time: we can wait a little while and things will work themselves out.

The cortex on the other hand thinks long term. In a practical sense, this means planning what you want to do with your life in the next twenty years, thinking about what kind of country you want to settle in, how many children you want and what kind of career you want to develop. It's about remembering the past and thinking about the future.[4]

"This isn't the meeting. This is the pre-*pre*meeting to talk about when to meet and *plan* the meeting."

The Japanese concept of time is very cortex – thinking in terms of a continuum. Before making any decisions, the Japanese think of posterity, honouring their ancestors and honouring the future descendants who will carry on their family legacy.

Space

When we look at space, things get a little more interesting.

If we look at the diagram on p. 60, we can see that, in reptilian space, everything that goes in must come out. Whatever you consume or inhale must be excreted or exhaled. It's very simple and basic.

The limbic space lies in between the other two. It's about waiting and delay: I may be waiting for you and you may be waiting for me, so we are in that space in between, the limbic space, a delayed dimension. Levi Strauss used to say, 'A mother is not a woman, a mother is a space between a woman and a child.' It is the space in between, like Limbo.

The space for the cortex is long-term planning. But the space dimension for men is different from that for women. Men's cortex space dimension is about being better, faster, stronger: always wanting more. Whereas the cortex space dimension for women is about cycles.

When we differentiate interior design from architecture, we are talking about the difference between the reptilian and the cortex spaces. Interior design involves making the inside pleasant, appealing, aesthetic and comfortable: it's about satisfying our direct reptilian needs. Interior design is the realm of women, who want to make the home not only safe and cosy, but also pleasing to the eye. Aesthetics are very reptilian because they deal with pleasure and instant gratification.

Architecture on the other hand is much more cortex: it's about making a structure functional and obeying the rules and logic of construction. It's a way of making a statement about a man's territory and power. It requires a higher knowledge of mathematics, physics and drafting. Many architects must even be well versed in local and state building codes, regulations and permits. This is the cortex at work: planning and logical thought.[5]

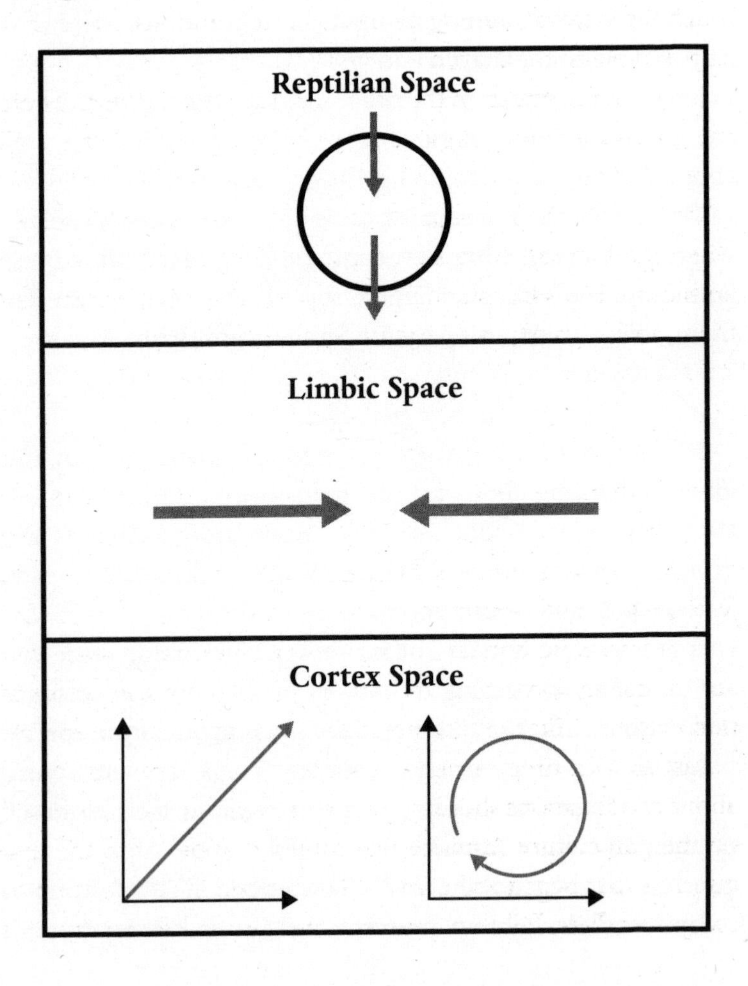

Now, if we apply this understanding of space to places of worship, we find incredible differences. Catholic churches and cathedrals are fascinating because they are often large with high ceilings and are elaborately decorated as a way of expressing God's majesty. Often centred on a statue of Jesus crucified on the cross, the bleakness and suffering is contrasted with images of angels, sun-kissed clouds and hopes of what heaven might look like. Like the reptilian, churches are spaces that touch the visceral, aiming straight at our emotions, expressing pain and pleasure, heaven and hell.[6]

Jewish synagogues on the other hand are very austere. There are no distractions, religious icons or imagery; rather it's all about abstraction and thinking: they are spaces for learning and gathering. For the Jewish faith, life is about knowledge, because when the Messiah comes everyone will be judged. This is the limbic space, it's like standing in a lift. You are neither here nor there, so you must wait patiently in the limbic space.

Energy

If we are talking about energy, the reptilian shouts 'intense!' It's a reflex – just as when someone grabs you by the shoulder as you're walking home alone late at night and your reflex is an intense jolt and sometimes an aimless swing. The limbic energy is to wait: it means being patient and waiting while you are expecting something to happen. Energy for the cortex is non-existent. Like the Internet, the cortex requires little energy.

Just as with time, American energy is very reptilian. It's all about action, about shooting first and thinking later. It's based on the gun culture America was founded upon, from the first gunshot that began the race for land in the Wild West, up to today's endless fight to preserve the Second Amendment's right to bear arms. America has a reptilian energy, and Nike's famous advertising campaign says it all: 'Just do it.' Don't think twice, just make it happen![7] 'Why would you do that?' someone may ask, and the response is, 'Because I can!' Americans do not need an explanation or a reason to do something.

As mentioned previously, for many Latin American cultures time is very limbic. It is therefore not surprising that energy is also very limbic for these countries. In Spain it is common to take a midday nap, or siesta, a moment to relax after working and before enjoying a nice meal with friends and family. A famous Mexican song says, 'Two steps forward, one step back' – limbic energy. If we look at Mexico's energy industry, for instance, until recently it used to be completely in limbo, neither private nor public. This hindered its development, because Mexico's limbic energy was not entirely a market economy nor was it fully state-owned either. In 1938, the government seized control over the oil industry, but it was not until 2013 that President Enrique Peña Nieto was able to achieve what had looked impossible for almost seventy years: to enact an energy reform that clarified property rights and would enable competition and investment. This is a truly historic reform in a country where the energy industry had been a cultural impediment to development.

French energy, on the other hand, is much more cortex. It's slow, epistemological, it's about thinking first and acting later. 'I think, therefore I am.' The French put all their energy into thinking, taking their time. In France, dining is a slow process; and medicine is an art, not a science. Their energy is focused on honing a skill, on talent. The word 'chef' is used to refer to not only the master of the kitchen, but also the orchestra conductor.

We've seen which cultures lean towards the reptilian, limbic or cortex when it comes to how time, space and energy are interpreted. But the layers of understanding are endless, and it's impossible to say a culture with a reptilian energy is better than one with a cortex time; or if that, in turn, is better than one with a limbic space.

Rather, interpreting the three brains and the time, space and energy dimensions help us decode cultures and comprehend which ones may allow us move up. That's what the next chapter is for. We are going to explore the ideal scenario, and, as you can probably already imagine, it all goes back to the reptilian, which always wins.

4 *The Ideal Scenario*

Don't Stop Moving

From the evolution of species to the evolution of a lifetime, we are always moving up. When recovering in hospital, one of the first things we are told to do is to stand up and go to the bathroom, walk around or simply move. In the morning we have to get up. If you believe in the evolution of species, life began under water, then we crawled on to the shore, grew legs and stood up. Since the beginning of time, we've been programmed to move – from Icarus trying to reach the gods, all the way to putting the first man on the moon, we've continually tried to move up, up, up.

If you don't move, you're dead. The air in your lungs and the blood in your heart need to move continuously. When you consume energy in the form of food, it moves its way through your body to nourish you and keep you alive. If you are reading this book right now, you're already one step ahead of the game.

Like Spencer's survival of the fittest, cultures too must continually compete if they expect to move up. Tokyo, Dubai and Kuala Lumpur have all been competing for the tallest building in the world: they all want to move up and *show* that they are moving up. You could say that this is very macho competitiveness, like teenage boys in the locker room comparing penis sizes. But for them, it's proof, a demonstration to the world that their culture is moving up. Look out.[1]

*

If cultures want to move up, they need to look to history, dig into their past and learn from it. History is our clue for understanding a culture. In 1803 the United States bought Louisiana from Napoleon, in 1845 they bought Texas from Mexico, and in 1867 Alaska from Russia. If the US wants something, they buy it, and that is their culture: a consumer culture.

We could learn a lot from the reptilian game of wars and conquering territory. Invasions tell us about a game of power and who held it at a certain period in time. The Roman Empire expanded across the furthest reaches of Europe and included parts of Africa and the Middle East; the Mongol Empire took over most of Asia; and in modern history the British, Spanish, French, Portuguese, Italian and German empires touched every continent on the planet.

European history has been wrought by invasions. The end result is ultimately devastating and detrimental to people's survival. During World War II alone, the number of deaths is believed to have been between 50 and 70 million. Though a disaster, the war brought about a shift in the way Europe seeks survival: to cooperation. In order to survive, the most powerful European countries decided to integrate and create a European culture. Though imperfect, so far it has worked. The European Union is a success story: it has managed to avoid war despite the economic crisis of the twenty-first century.[2]

Striving for survival is the reptilian at work, but making a conscious agreement of cooperation and integration is the cortex. This is the cortex comprehending the reptilian: it's the ideal situation. When cultures comprehend the reptilian and use the cortex to take it to a new level, what follows is prosperity and moving up.

Old Solutions, New Problems

A long time ago, a tribal leader decided to cross the desert and attack another tribe that was a two-week walk away. The warriors prepared their weapons and collected their food, especially their favourite food, meerkats. After several days of walking, the chief realized that some of his warriors were getting sick, and deduced that it must have been due to the old meerkat meat (rather than the heat, mosquitoes, and so on). The leader then decided to prohibit the tribe from eating meerkat meat. When the chief died, his son declared meerkat meat to be taboo; and then when he died, his son declared eating meerkat meat was against God's will and whoever deviated would be damned to hell.

The tribe still lives in the desert to this day, but each year many die from malnutrition even though they live surrounded by meerkats. Today, new techniques have been developed that could allow them to preserve meerkat meat, making it safe for consumption, but unfortunately for them the law of God is immutable.

Crystallization occurs when a solution that was good at a certain time for survival turns into a rigid law that is disconnected with today's reality. If a culture does not allow one to question the law, the old solution might become a 'mobility killer', and bring the culture to its end.

Hundreds of years ago, the Indian countryside relied heavily on dairy products from cattle, so much so that the Hindu religion prohibits practitioners from eating beef. Now, many die of hunger in India, in an environment where cows are plentiful. This can be seen as a law that has been crystallized by religion and therefore remains unquestionable.

Technology has provided us with a variety of contraceptive

options that were unavailable a century ago. But while technology has provided women with greater freedom of choice, many cultures have failed to adapt to it. Taboos against female promiscuity have already been crystallized in many cultures.

The concept of karma is a very romantic notion based on the law of attraction, where like attracts like and for every action there is a reaction. Some Asian religions have adopted the law of karma in their teachings – an efficient way of ensuring good behaviour. But when karma is used to justify India's caste system, we have more negative consequences. Karma has become so crystallized in Indian culture that it affects average people's ability to move up.[3]

One more way of hampering movement is through the law of reversal, when something that was good at a certain time or in a certain society is no longer applicable. This is the case when

Old solutions, new problems

we think about powerful unions, for instance. Trade unions were created to defend workers' rights, a just and noble cause. But today, too often unions are found to be not only corrupt and abusive of their power, they are also job and quality killers.

The law of reversal doesn't require a rigid law to be passed, just that an idea is believed to work anachronistically or independently of its environment. When we in the west talk of child labour, we are quick to judge and create international laws that discourage the practice. But the law of reversal shows that in some countries, if children don't work, they don't eat. Laws can be OK in one environment, but have negative consequences in another.

The law of reversal and crystallization show that a culture that fails to self-consciously and dynamically adapt will fail to move up.[4] It's basic survival of the fittest: closed systems which have old customs that are crystallized, or which hang on to old solutions that are no longer useful, are not going to move up – they may even die out. On the other hand, open systems which are constantly changing, adapting to new challenges and respecting the reptilian are going to move up.

Enough Is Not Enough

> But what is happiness except the simple
> harmony between a man and the life he leads?
>
> Albert Camus[5]

So if our biology programs us to survive, and our bio-logical forces are what drive us (reptilian always wins, remember?), shouldn't it be enough to be biologically programmed to survive, as we are, so that everything adjusts itself automatically?

Well, no. Biology is enough to feed our desire to survive and reproduce, but it's not enough to actually enable us to survive.

Where does our culture come in, then? Biology and culture interact in intricate ways, sometimes handicapping our ability to move up through the law of reversal or crystallization. How can culture help our biology move up?

The ideal scenario is not a recipe with ingredients and methodology to amend a culture. If this were the case, the ideal would be to create cultures that are the same. Rather, the ideal situation is that cultures accept the three parts of their brains, *especially* the reptilian. But not only that; the ideal is that our cultures in fact turn our reptilian desires into pleasures with the help of the cortex.

Starbucks understand the cortex: their coffee shops are comfortable, the service is pleasant and quick, and you don't have to walk very far to find one. But what did Starbucks do for the reptilian? They offered 87,000 combinations of beverages.[6] At Starbucks, you don't just buy coffee, you also buy identity. They write your name on the plastic cup. They call you by your name, and the barista smiles at you. They make you feel special and unique. What Starbucks is selling, and what the reptile inside you is buying, is social recognition.

Why is it important for cultures to turn our reptilian desires into pleasures? Because pleasure is crucial for survival. When we feel rewarded for eating, sleeping, having sex, and other reptilian behaviours, we are more likely to do them again and do them well because it benefits us to do so. Most cultural institutions are deeply concerned with controlling pleasure, creating rules and codes surrounding which pleasures are permitted and which are not, and how we are supposed to behave with regard to these pleasures. As we've already said, the reptilian always wins: it's the most important part of survival and reproducing our genes – without it we'd have no way of moving up. So the most successful cultures are those that turn reptilian desires into pleasures.

The purpose of a culture is to relieve your anxiety and guilt. Cultures are meant to give us a reference system of behaviour in order to make us feel comfortable. The purpose is to make you feel like an insider, like you are in the group, following the rules and feel a sense of belonging; like you are doing things right. That is survival, because if you are out of the group you have less chance of surviving.

Denying the biological characteristics of humans won't make them go away; so the best thing to do is to accept them, assimilate them and act with the knowledge that they are there. The Muria people live in central India. Unlike many western societies that shun adolescent sex, the Muria accept that children are sexual beings from birth and even reserve a physical space for them to explore their sexuality. This place is called the *ghotul*, a house where adolescents of both genders can begin to experiment with premarital sex, and even learn how to enjoy it. It's an initiation. Having sex is something everyone has to learn how to do at some point, and what better way to learn than by celebrating it in a safe and open place that is guilt-free?

The same thing happens with food. In France, the pleasure of eating food is an art. If you don't have the time to enjoy it, you should skip it. The French are famous for cuisine and the rituals surrounding it: which wine to have with which dish, what spirits are better before or after dinner. In Italy, eating is an involved process and cannot be accomplished in half an hour. Food for the Italians is an excuse to get together, to socialize with friends and family.

In France, it is common for under-age children to begin drinking alcohol, tasting wine and spirits with their parents and honing their taste for them, whereas in the United States children aren't taught to appreciate alcohol – rather they perceive it as a means of getting drunk. Eating and drinking is a basic reptilian necessity, but French culture has learned the art of pleasure surrounding food.

Some cultures are very good at taking care of the reptilian. It was a Roman who said, '*mens sana in corpore sano*'; your mind and your body should be kept healthy. Cultures differ in the ways they deal with body development, food, sex and desire.

You can also tell a lot about a culture by the language they use. In English you say 'thank you' and in German '*danke schön*', and both have their roots in the word 'think' – very cortex. In France, '*merci*' is linked to 'mercy' and 'merit', between emotions (the reptilian) and the rational (the cortex), so it's very limbic. Italian '*grazie*' and Spanish '*gracias*' are linked to the grace of Christ and illumination: you are special (remember Starbucks?), back to the reptilian.

In many northern countries, where the population spends over six months without light during harsh winters, cultures have no option but to adapt to the environment. Through our research, we have found that many of these cultures are rich in creativity. With no other option but to stay indoors, they need to get creative and find new and interesting ways to entertain themselves. It is in their basements that Scandinavians create new music, literature and art, not to mention much more sexually liberal ideas.

German culture is famous for its strong cortex dimension, always creating a system, a process that works through logic and reason, founded on science and technology. The Germans aren't famous for their cuisine, or for their romantic courtship rituals, but they are famous for their cars. The best cars in the world are German: Porsche, Volkswagen, Mercedes and BMW are brands we all recognize.

We have found that the ideal scenario is a sort of harmony between the reptilian and the culture, and where the cortex is also involved. Cultures must always be thinking about reptilian needs in order to properly move on to the needs of the limbic and the cortex, those higher pleasures. You can't go to the opera if you're not breathing. But, in order to do this, cultures need to adapt, to change with the changing environment, and those that fail to do so will be left behind. In essence, cultures need to keep moving.

C^2 = CULTURE CODES

5 The Five Critical Moves

Get Out the Board Game

High up in the mountains of Tibet, a group of young apprentices were learning how to master the game of chess from one of the greatest chess teachers in the world. The master told his students, 'In chess, even a piece of the lowest rank, a pawn, can reach the top and become a queen.'

The students were puzzled. A young British boy said, 'In my country, a queen is a queen, and a pawn is a pawn.' A South African girl blurted, 'In my country, it is but a dream for a pawn to become a queen.' And a boy from India said, 'In my country, it is a sin for a pawn to become a queen.'

The master shook his head and told his students, 'The rules of chess are unquestionable. It *is* completely possible for a pawn to become a queen. All you need is a good strategy.' The game of chess is about moving up. A country that allows a pawn to become a queen is a country that is moving up. What we want is to move up, that is our strategy; and the 'five critical moves' are the tactics that will implement this strategy. These tactics are also the basis to decode the 'C^2' (culture code) value for our index. These are qualitative considerations that, along with the 'third unconscious', which we will analyse in the following chapter, will enable us to determine the C^2 values for seventy-one countries according to their ability to move up.

1. *The Game Is Not Always What It Seems (Below the Sea Surface)*

Especially for non-pros, the tricks of the game are not always apparent and can only be discovered over time and with experience. And, just as is in any game, not all aspects of a culture are instantly apparent or obvious to the external observer, and are sometimes even less noticeable to the insider.

When flying, you may catch yourself gazing out of the window, fascinated – perhaps even entranced – by the vast ocean below. But what you see makes up only approximately 5 per cent of what's really going on in the ocean; the other 95 per cent is what lies beneath the surface. Similarly, it is below the surface that we find all those aspects of our cultures that are not immediately obvious, such as how to behave when meeting the in-laws. The deeper you go, the more important the cultural elements become.

Actually, the bottom of the ocean is the foundation for what you see at the top. If you want to understand the underlying reasons why people behave the way they do, really getting to the heart of a culture requires seeing past the surface. The deep sea is where you are more likely to find the answers about what makes people move up.

The sea's surface is made up of stereotypes: they are often superficial and ignore what's really driving the culture. They involve making assumptions about a culture based only on the behaviour of a few individuals. Making cultural generalizations, on the other hand, allows us to observe the general tendencies of a group's behaviour – its traits, preferences and so on – and analyses them empirically through qualitative and quantitative data.

There's an old joke that goes: paradise in Europe would be

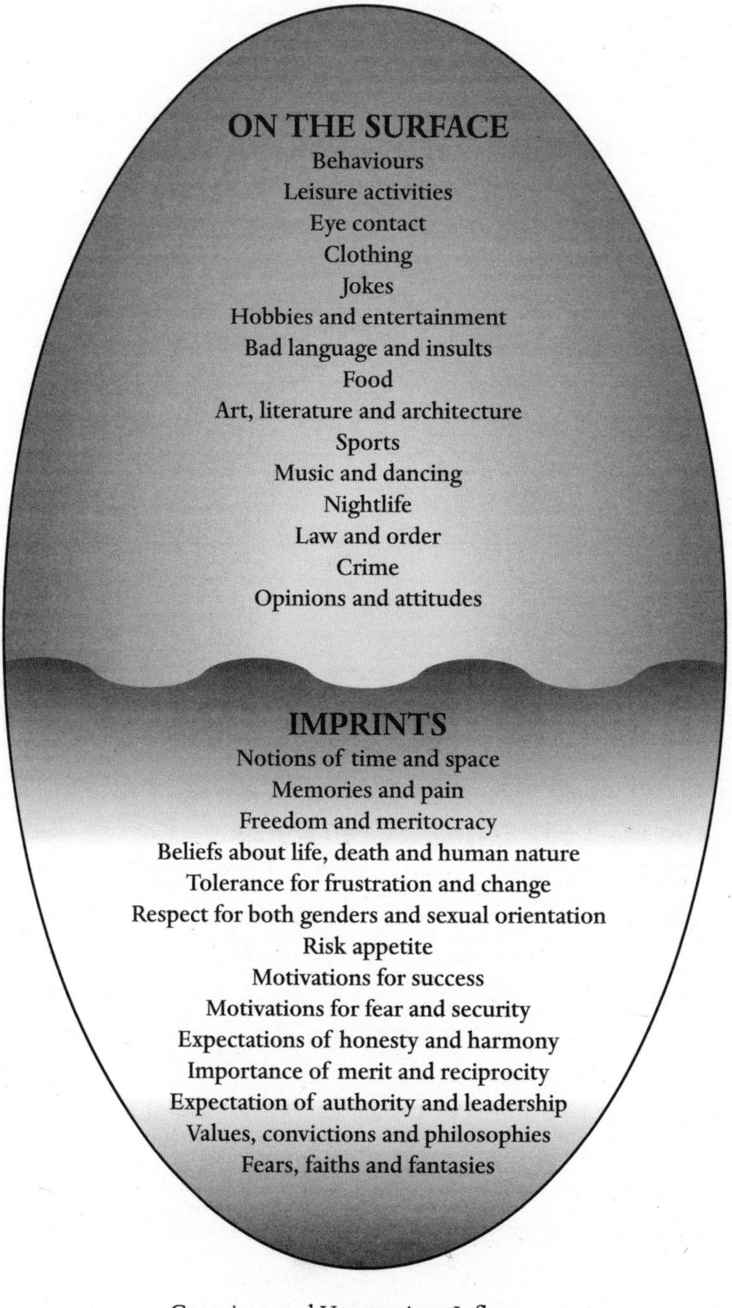

Conscious and Unconscious Influences

Would you consider these stereotypes or generalizations?

if we could find a chef from France, a policeman from Britain, a mechanic from Germany, a lover from Italy and an administrator from Switzerland. On the other hand, hell would be if we got together a mechanic from France, a policeman from Germany, a chef from Britain, an administrator from Italy and a lover from Switzerland.

Though humorous, stereotypes frequently fall short. Some of the best restaurants in the world are found in Britain and some of the most desirable men can be found in Switzerland.

A stereotype is focused on individuals, analysing the behaviour of a few people and ascribing it to the whole group. In contrast, cultural generalizations are group-focused, analysing the culture as a whole and then applying those insights to the individual.

Cultural generalizations that come from research and from insights by informed international cultural experts allow us to paint a fairly accurate picture of how people in a given country are likely to behave. This is an essential tactic when determining which aspects of a culture make people move and which inhibit movement, which we will look at later when we discuss the C^2 variable and our culture codes.

2. Call My Bluff

You can't always believe what people say.[1] Your opponent wants to win just as much as you do, and they will do anything in order to succeed. People will try to build conceptual maps, processes, formulas or strategies to cope with certain obstacles – they will try to be entirely rational, but let's remember that emotions drive our decisions and determine what we want in any given situation. People's true intentions, feelings or perceptions are not always apparent at an initial glance because they cannot be rationalized instantly and need to be analysed in more detail.

What does success mean to Americans? Most would answer: honest hard work, good luck, good health, good genes. But we don't believe them. What about money, power, big houses, fast cars and romantic success? The most important thing to bear in mind when interpreting culture codes is that you can't always believe what people say. In a traditional focus group, interviewers ask direct questions. They question people about their preferences; and those interviewed tend to give the answers they believe the interviewers want to hear. We are not suggesting that people lie intentionally. This happens because people tend to answer questions using their cortices, the most rational part of their brains: the part that controls intelligence rather than emotion or instinct. And in many cases the cortex doesn't really know why we do the things we do. So people believe that they are telling the truth. A lie detector test will confirm this. In most cases, however, they aren't saying what they truly mean.

In elections people are often surprised by the results, since the polls predicted something entirely different. The reason is precisely this: the way in which questions are asked and interpreted by people does not reflect what individuals truly mean. Even the most self-aware are rarely in close contact with their unconscious. We have little interaction with this powerful force that drives so many of our actions. Therefore, we give answers to questions that sound logical and are even what the interviewer expected, but which don't reveal the unconscious forces that determine our feelings.

So in order to get to what people really mean, to find out what really motivates us, we need to tap into our emotions and our instincts: our unconscious.

3. Once It's Written, You Can't Change the Rulebook

In *The Culture Code*, Dr Rapaille writes, 'There is a window in time for imprinting; and the meaning of the imprint varies from one culture to another.'[2]

The reptilian brain generates instinctive behaviours that help us survive, instincts that all animals, including us, have in common. Any living being that has survived up until this point is already winning the game; the losers are those that are now extinct. The American philosopher and psychologist William James defined 'instinct' as 'the faculty of acting in such a way as to produce certain ends, without foresight of the ends, and without previous education in the performance'.[3] We all have instincts and we express them in diverse and creative ways, but essentially they are our primary drives that push us to survive, to win.

Fear is one of our most natural instincts. It has helped us survive thousands of years of history. Think about the animal you fear the most. You can probably clearly remember the first experience that triggered that fear; you may also have dreamed about that animal; and your fear may even have transformed into a phobia.[4] Nevertheless, whilst our fears are mostly irrational, they somehow manage to survive. That's because our emotions are the most powerful influences on our memory, which is why our cultural imprints are firmly linked to our emotions and perceptions. My imprint for a dog may be the time the neighbour's dog bit me when I was four years old.

Emotions are the keys to learning, and to imprinting. The stronger the emotion, the more clearly the experience is learned. They are mental highways that are reinforced by repetition. In the same way as language is learned and acknowledged by its usage and not by mere definitions, emotions which are experienced frequently will soon be associated with certain feelings and therefore with certain persons or situations. If your mother makes you a peanut butter and jelly sandwich every day for lunch, you will begin to associate peanut butter and jelly sandwiches with maternal love.

We are most susceptible to imprinting before the age of seven. This is the critical period when most of our culture's imprints are placed within our subconscious, and they are usually defined by the culture in which we are raised. For instance, a child raised in the US experiences their most active learning period in an American context. Therefore, all of the child's cultural imprints and references are created within this American context, making the child an American, rather than a Kenyan.

Because our strongest imprints occur at a young age, we never get a second chance at acquiring truly significant imprints from another culture. If your parents are Russian but you spend most of your formative years in Argentina, most of your cultural codes are likely to be Argentine.

In order to understand the culture code for different attitudes, objects, concepts or emotions, it is important to tap into those early imprints. If you want to find out what 'career' means to a specific society, you don't just outright ask people what it means to them, but rather what is their first memory of the concept 'career'.

4. Levelling the Playing Field

A fair competition guarantees the best winner. In most board games, like Monopoly or Snakes and Ladders, all the players start from the same place. No one gets to begin ten points

Social Mobility *by Danish and Norwegian duo Michael Elmgreen and Ingar Dragset*

ahead. This is because games have rules, and the rules apply to everyone, ensuring a fair competition. The same goes for outdoor games: a level playing field ensures the most efficient allocation of talent, and the best player is sure to win.

Depending on when and where you were born, you will have different opportunities for moving. In one place, a lower caste Indian might not have a chance of becoming a Brahman during their lifetime, whereas in another the members of The Beatles could be honoured by the Queen in just a few years.

But for the most part we all want to move up. It's not just about jobs and money, or security and freedom, religion and culture, or laws and regulations. Moving up is all those things and more. People want to move up for a variety of reasons: for status, the opportunity to have children, money, fame, happiness, love, stability, 'immortality' – anything, really.

'You can be anything you want to be' is a well-known phrase that not enough children around the world hear, one that isn't true for most people. The concept of equal opportunity and life chances goes hand in hand with what this book proposes: we all want to move up, and hence we should all have the opportunity to move up. Why? Because we matter.

For this reason, meritocracy is an important tactic in the strategy of moving up. If we all want to move up then we should all be able to move up. The way in which a society moves is a basic factor in determining if it is moving up or down. If our society decides who is at the top by factors such as being well connected to those in power – wealth, nepotism and so forth – then just a few will be able to move up. Even worse, those who are on top will maintain their position for generations, stopping the natural flow of social mobility.

Meritocracy changes the ways in which we can move up.

Status is not defined by *who* you know, but *what* you know. Individual talent becomes the defining factor that determines who has the chance to be on top.

5. The Double

We are not blank slates. Philosophers, theorists and scientists have been pondering over the nature / nurture debate throughout history. In the nineteenth century the scale was tilted towards nurture: the human mind as a *tabula rasa*, a blank slate waiting to be moulded by our experiences and environment. There is no denying that our environment has a significant impact on our individual behaviour and personality, and these differences can also be observed in the rich variety of cultures that human beings have developed throughout the globe. Even so, there are approximately 370 universal traits shared between all cultures, as identified by the anthropologist Donald Brown, who calls them 'human universals'.[5] Although each culture expresses these universals in different ways, every culture attaches some kind of meaning to concepts such as death, beauty, status, aesthetics, leadership, personal names, reciprocity and sanctions.

Since the nineteenth century there have been remarkable scientific discoveries that have led to an understanding of the interplay of nature and nurture: genes do not directly create a behaviour, and the direction of causation can go both ways. The relationship between biology (nature) and culture (nurture) is a symbiotic, interactive and motivating process, rather than a competition. The 'blank slate' is improbable and babies do not emerge as personality-less zombies, as anyone who has reared a child will know.[6]

So If You Want to Play . . .

and not only play but win, don't forget these five critical moves:

1. We need to look to the deep sea for answers; the surface is not enough.
2. If you want to understand imprints, don't believe everything people tell you.
3. Imprints are complex and vary from culture to culture.
4. Meritocracy should determine the winners.
5. Nature affects nurture, and vice versa.

6 The Third Unconscious

Don't Believe Everything You Think

Every extension of knowledge arises from
making the conscious the unconscious.

Friedrich Nietzsche[1]

The job of the C², or culture code, is to help us understand who we really are, below the sea surface. According to Dr Rapaille's method, in order to determine a value for C² we must understand how cultures work through the unconscious and which specific cultural traits favour mobility.

Because of his quest for survival, man has been forced to keep moving since the beginning of time. As mankind developed we began to settle permanently in small areas, then in villages and finally in big cities and even countries. By living together day after day with the same people in the same places, we moved from merely sharing genes for survival to sharing knowledge, tastes, traditions and beliefs which have given us a collective identity far beyond just being human. The human race began to generate some kind of culture. This culture finally gave us what we might call 'a place in common with other people'.

But let's take a step back: what is the significance of man's ability to settle in one place? The obvious answer would be to say that we found the necessary technologies (like agriculture) that could sustain us in that place for a long time. This shift created a radical difference between humans and other animals: human beings were able to create a system that went beyond what was given to them by nature. How can we explain this change? Thanks to a great science called evolutionary psychology we can provide a proximate answer.[2]

The human species, as we know, has evolved not only physically but also psychologically. Throughout thousands of years, man has been perfecting his species through adaptability but there was a moment in that evolution where the changes made the difference: at some point in this evolution man found himself. This is to say that we became conscious, and with this we were finally able to understand ourselves as a species capable of changing our environment and, perhaps most importantly, changing ourselves. We began thinking about ourselves not only as bodies with needs and instincts, but as individuals with thoughts and choices. Societies and scholars have typically centred their studies on the human unconscious, an unconscious that has made us behave in ways we don't understand and which varies greatly across individuals.

The first level of consciousness is the *individual unconscious*, a concept inspired by Sigmund Freud, whose life's work centred on the unconscious. At the end of the nineteenth century, Freud travelled to France to work with Jean-Martin Charcot at the Salpêtrière Hospital in Paris to study the unconscious in cases of hysteria. Freud was able to develop our first understanding of the unconscious forces that make us do things that we don't fully understand.[3]

Throughout most of history, it was thought that our conscious self was most dominant, but in reality our unconscious arose first in evolutionary time. And various unconscious systems have proven themselves useful in guiding our different cultures. This allows us to understand new scientific discoveries that highlight the complexities of the unconscious – for instance, how our neurons can make a decision in the ten seconds before an individual carries something out.

Though avant-garde, Freud's work was very macho, focusing on the hysteria of women, not men. The unconscious Freud analysed was very much in sync with what we understand today as the reptilian. For Freud, the rules of our society repress the unconscious. He believed for instance that children unconsciously have a possessive and even sexual obsession with the parent of the opposite sex, called the Oedipus and Electra complexes (obviously, these ideas are considered taboo in most societies and are therefore repressed by these cultures). Today, we would say that this is the reptilian being repressed.

Then we have the *collective unconscious*. Carl Jung developed this concept through his work on archetypes.[4] We all have an archetype for 'mother' – a concept of what a mother is and what she represents – though one individual may associate mothers with power and discipline, while for someone else they are linked to unconditional love and affection. But the point is that we all have a concept of mother, just as we all have a notion of death, birth, puberty, reciprocity, courage, violence and so forth.

When we talk about love, we are talking about a universal concept: it's part of the collective unconscious. But all over the world, finding a perfect mate is one of the biggest challenges people face. That's why the American anthropologist Dr Helen Fisher created archetypes of love chemistry. On the website Dr Fisher advises, chemistry.com, people can create a profile and the website will match them with potential mates from an anthropological point of view.[5]

Freud said that biology is destiny. Jung looked at how this common biological structure (for example, male, female, life, death, growth, ageing) is expressed through universal archetypes such as the mother, the woman and so on.[6] Both men developed provocative and insightful new theories, but psychology, specifically evolutionary psychology, has changed a lot since then. We've discovered new ways of analysing sexuality, the unconscious and so forth; but what we are really interested in is discovering how the cultural unconscious is organizing our behaviour.

The Ghost at Work

The third unconscious is what we are most interested in: the *cultural unconscious*. Freud and Jung approached it in their work but never really got close enough. Our cultural unconscious is the way each culture deals with the tension between the cortex and the reptilian.

We are among the first to develop the concept of the third (or cultural) unconscious. Adolf Bastian pioneered the idea of *Volksgedanken*, or 'ethnic ideas explaining universal archetypes'. We call these 'reptilian structures'; they are universal schemes or schemata. Each culture functions as a survival kit inherited at birth, which transmits 'solutions' to help us deal with biological needs. We call these sets of forces the 'cultural archetype'.

Our best tool for understanding the cultural unconscious is through culture codes. The culture code 'is the unconscious meaning we apply to any given thing via the culture in which we are raised'.[7] Because our cultures evolved differently and are expressed differently, on an unconscious level our culture codes are different as well and lead us to process the same information in different ways. These hidden codes are what will help us evaluate the C^2 value for our sample of seventy-one countries.

Each culture has a different code for different things, a different cultural consciousness, whether it be a code for a certain concept (such as puberty or courtship), an object (such as cars or food), or an emotion (love or anger).

For example, cleanliness is a universal concept, but how does a particular culture perceive cleanliness? The code for clean in Japan is different from that in China. Germans are very clean, the French are not; Japanese are very clean and the Chinese are not. In a study we did in France, we found that only 1.5 million toothbrushes were sold in a year to a population of 55 million. That says something about their code for cleanliness.

Culture codes are often hidden and unconscious, and they therefore require a very precise method for decoding (which

we already touched in our chapter about the five critical moves). To uncover a given culture's code, we must first analyse the imprints that make up the code. An imprint is the combination of an experience and its accompanying emotion.

As French scientist Henri Laborit emphasized, there is a strong connection between learning and emotion.[8]

Fear is one of our most powerful emotions, so if as a child every time a dog approached you your mother grabbed you harder and expressed her terror openly, you would naturally have a very strong imprint for dog that was associated with fear. If in your last few relationships your partners cheated on you with Lebanese women, you will tend to have a strong imprint for Lebanese women associated with jealousy. Once an imprint occurs, it strongly conditions our thought processes and shapes our future actions. Imprints vary from culture to culture and from individual to individual, and early imprinting has a tremendous impact on why people do what they do.

An imprint and its code are like a lock and its combination. If you have all the right numbers in the right sequence, you

"Today: The collective unconscious..."

can open the lock. When all the different codes for all the different imprints are put together, they create a reference system that people living within a culture use without even being aware of it. This reference system guides different cultures in very different ways. For instance, many European countries have liberal sexual attitudes, so, when deciding whether to have a one-night stand or not, if raised in Europe you will refer back to your reference system and most likely decide that it is indeed a socially acceptable behaviour. Therefore you have fewer qualms about acting on your sexual desires. We can therefore deduce that most Europeans would be more open to having one-night stands compared to other cultures.

Cultures are created and evolve over time, though the rate of change is glacial. When cultures *do* change, the changes occur in the same way as in our brains – via powerful imprints. These powerful imprints alter the reference system of the culture, and the effect trickles down to subsequent generations.

World War II created powerful imprints for all Europeans, whether for the immediate victims, the perpetrators, relatives, bystanders or descendants; and each country's imprints varied greatly, having a direct effect on the change in reference systems. Before the war, Germans probably had a completely different culture code when it came to being openly racist or anti-Semitic; whereas now, it is unthinkable to make racist remarks in public due to the stigma surrounding German intolerance during the war. For many Jewish people, the culture code for Europe too has a different meaning after the Holocaust.

As is evident in the previous examples, there are some things that we don't perceive in the same way as someone raised in a culture different to ours, and this collective unconscious may be captured in a culture code. Ideally, culture codes understand the reptilian dimension of our cultures: how some cultures are more in tune with the reptilian, while others are more repressive.

Expression v. Repression

Like the current of a river pressing against a dam, our reptilian puts pressure on us, forcing us to act on our instincts to eat, sleep, have sex and so forth, and to act on them now. When we are able to act on this reptilian pressure, we are relieved. But when our culture inhibits us from expressing our reptilian properly, there are a variety of negative consequences.

Let's take napping as an example. Taking a nap after eating is very pleasant and can make us more productive, since our body has dedicated some time solely to digestion. However, many of us simply cannot go to sleep after eating and must carry on working. Who is the culture deceiving? Your obligation to continue working after eating won't make your sleepiness go away.

If we look at the way pressure (*'pression'* in French) has been diagrammed on p. 98, we find there are several possible reactions: ex-pression, re-pression, and de-pression. The reptilian creates a pressure on a culture and, if the culture can properly express the reptilian, it is healthy. But if the reptilian represses itself it backfires in the culture's face, for instance women in Saudi Arabia who want to be able to drive.

If a culture doesn't allow the reptilian to express itself, pressure builds up, and there are two possible results: depression, where the culture feels a sense of hopelessness that no change is possible, which provokes a violent or greedy reaction; and repression. Japanese culture is very repressive of many reptilian needs, and the Japanese can also be very violent. During the Nanking Massacre in 1937, Japanese soldiers not only committed mass murder of an entire city, they also raped any survivors.

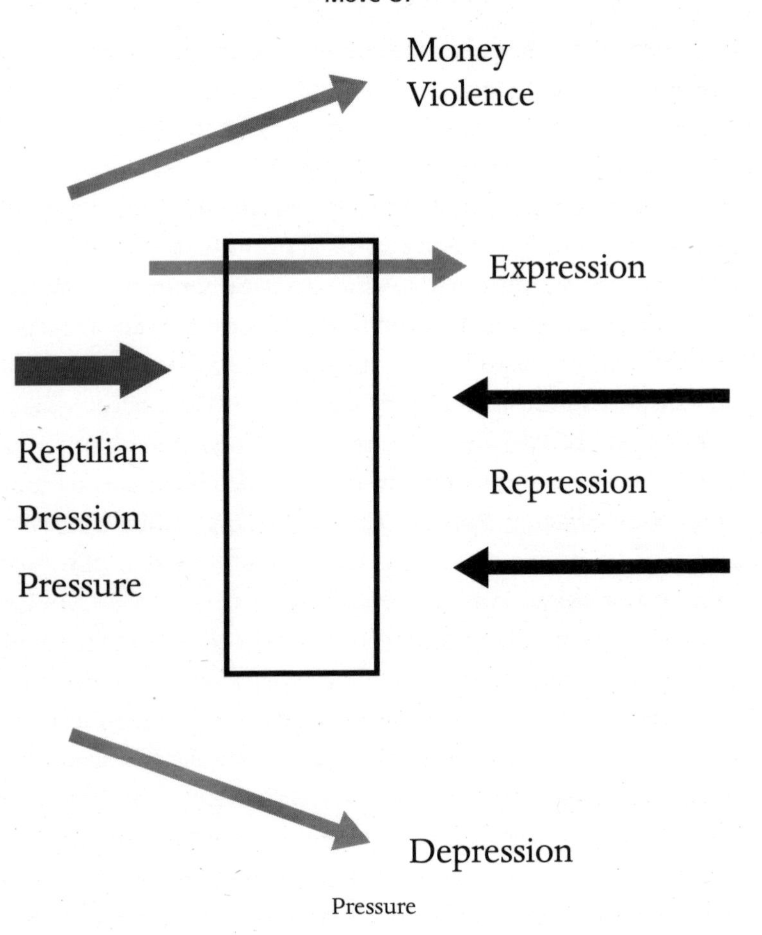

Pressure

Like Japan, Sudan has a highly repressive culture. Traditionally an Islamic country with rigorous and strict laws surrounding codes of conduct (especially relating to reptilian behaviours), Sudan has had the worst luck. We can link its highly repressive culture to the violence this African country has suffered for over a century.

Sometimes, cultures celebrate the reptilian. In France, the fact that couples are going to be attracted to more than just one person for the rest of their lives is openly accepted. So instead of placing high expectations on each other to be monogamous, they have the *cinq à sept* (five to seven) concept, when the couple are open to meeting their lovers on the side from 5 p.m. to 7 p.m. In America's puritan culture, sex should be restricted but food need not be.

Many monotheistic religions repress our natural drives, our reptilian, expecting us to control our desires for sex, drinking and eating. They teach us to deny our nature, have self-control and show restraint. But during the medieval era an incredible thing happened: junior clergy in France were given one day a year to entirely let go – one evening for complete debauchery and loss of control – the Feast of Fools. Clergymen could eat and drink uncontrollably, urinate and defecate at will, have sex with people from either gender, and could even dishonour the Bible on this special evening. The tradition didn't last very long but it was a way for a system of repression to allow some expression of the reptilian, knowing that expression relieves tension and that 'we cannot be naïve about our nature', as Alain de Botton puts it.[9]

In contrast, northern European cultures tend to be less repressive. If you look at the differences between the Protestant countries, such as Sweden and Denmark, and Catholic countries, such as Spain and Italy, there is a vast difference in repressiveness. It all comes down to how cultures deal with the reptilian, and then how the cortex comes into play and turns our reptilian instincts into cultural pleasures and art. We can hone our instincts so that we are not slaves to them, addicted to food, sex, power and ego. The cortex allows us to make best use of the reptilian, helping us to actualize our selves.

Be Aware

Now that we've understood our three unconscious levels, let's talk about what we can do with them: we can become aware. Awareness gives you peace of mind. First you have to be aware of your reptilian dimension – that you personally, collectively and culturally have needs that are beyond your control. There is nothing wrong with wanting to eat, or to have intimacy, or to have a drink. For every human being individual consciousness is unique; it is like a different language for every person. You are unique and have your own script. You can learn about patterns in your own life, habits and tendencies that you create. Elizabeth Taylor was an expert in divorce, not in marriage – that was her pattern. We have an unconscious way of repeating things, an unconscious set of patterns.

Biology determines who you are, but it does not dictate your final destiny: only you can decide that. It's your choice to what extent you accept your culture. If you are a woman born into a culture that doesn't suit you, it is incumbent on you to be aware of that and free yourself from shame or guilt. This freedom comes only if you are aware that culture is not destiny, it is a choice. You can choose which culture suits you most. To move up sometimes means to move away. You work out what doesn't suit you: that is about being aware of your reptilian basic needs.

In order to get to know your culture, you need to delve into the cultural unconscious, which is a mix of things: geography; history; your parents and how they transmitted their time, space and energy. These are all things that create unconscious codes, which are difficult to be aware of; but when we do become aware of them, we can make choices. Once we become conscious of those unconscious cultural elements, and make

"I hope you don't mind. He doesn't know he's a dog."

them available to the cortex, we get the 'wow!' factor. When we are aware of what was once hidden, we begin to say to ourselves, 'Wow – I already knew that all along!'

For instance, money in France is bad, money in America is good. In France, you are punished for having wealth in the form of higher taxes, and your status doesn't come from how much wealth you have and how you got there, but is measured by how you contribute to society. If you have a Rolls-Royce in France, they are more likely to say, 'One day, we are going to take it away from you.' In America they'll say, 'One day I'm going to have one too.' The cultural collective unconscious is shaped by narrative, stories, movies, clichés and stereotypes that reinforce the same structure. That's what the C^2 variable is all about.

For instance, to be neutral and peaceful in Switzerland

requires the Swiss to have a permanent army which is ready to fight at any time: a citizen's army whose members keep their weapons and ammunition at home. The Swiss undergo military training annually. They are strong and ready, and at the same time the army is where the different cantons and languages come together and create a national bond. It is astonishing that a neutral and peaceful culture is cemented by a very strong army.

So you can decide which culture suits you best by becoming aware not only of who you are, but also of what might help you, depending on your specific strengths and desires, to move up.

Mainland Chinese women might choose to go to Hong Kong to deliver their babies. Indian students might choose to go to London or Oxford in order to obtain a degree and learn how to speak good English. Most of the time, there is a set of elements that creates the foundation for success. Hong Kong has an incredibly business-friendly environment and children who grow up there have better chances of succeeding than anywhere else in China. London and Oxford have excellent educational systems and score very high on all four S's. This is because they respect archetypal values and the reptilian mind.

Former New York City Mayor Rudy Giuliani did something similar for his city. During the 1990s, New York was dirty and plagued by crime. He applied the broken windows theory, a theory that believes that urban disorder and vandalism have a domino effect on high crime rates within a city. If there are no more broken windows, then there is less crime. Rudy Giuliani made it his mission to clean up the city, having 'zero tolerance' for vandalism and crime, and applied this practice with the city's police department. Those caught were given no second chances and no mercy. Eventually, crime rates dropped and the city of New York turned into a safe and attractive destination. Giuliani once said, 'When you confront a problem, you begin to solve it.' As a political leader, he knew that if he let people urinate or drink on the street, the chain of consequences would be far more serious.

But in an interconnected world, where there are a million messages but the content is all jumbled up, it's difficult to find our priorities. We don't know what to do with all this information. The purpose of this book is to provide an understanding of how all this information needs to be organized and prioritized.

Taking all of this information into account, and most importantly organizing it, we can then determine which are the best places around the world for moving up. If for you moving up means having a big apartment, then you shouldn't move to London; if it means being in a place that is safe, clean and comfortable, you may chose Tokyo. If you want more than you have in front of you, you can either move cultures or do your best to transform your own.

BE BIO-LOGICAL

7 The Four S's

Since the beginning of time, humans have pondered one of the deepest questions that trouble us: What makes us move? In other words, why do some people strive to reconcile their biological battles and move up while others simply don't? Why do some countries grow and drive social mobility and others do not?

Maslow's theory is a good place to start answering this question, but it is not the ultimate answer. His powerful image of the 'pyramid of needs' has been one of the most cognitively contagious ideas in the social sciences.[1] Maslow felt obligated to develop this model of human motivation because he disagreed with the prevailing paradigm in psychology at the time, which tended to believe that all human motives could ultimately be reduced to a few primary drives, such as hunger and thirst. Social scientists at the time viewed people's desires for affection, esteem and self-actualization as secondary drives. Maslow did not buy the theory about secondary drives. He argued, 'We could never understand fully the need for love no matter how much we might know about the hunger drive.' He proposed instead that human beings have several completely independent sets of basic needs. If you have ever taken a course in organizational behaviour, or read a psychology book, you've probably seen the pyramid that represents his ideas.

At the base of his pyramid lie our immediate physiological needs, like hunger and thirst. They are our biological priorities, essential to our survival. But once our immediate

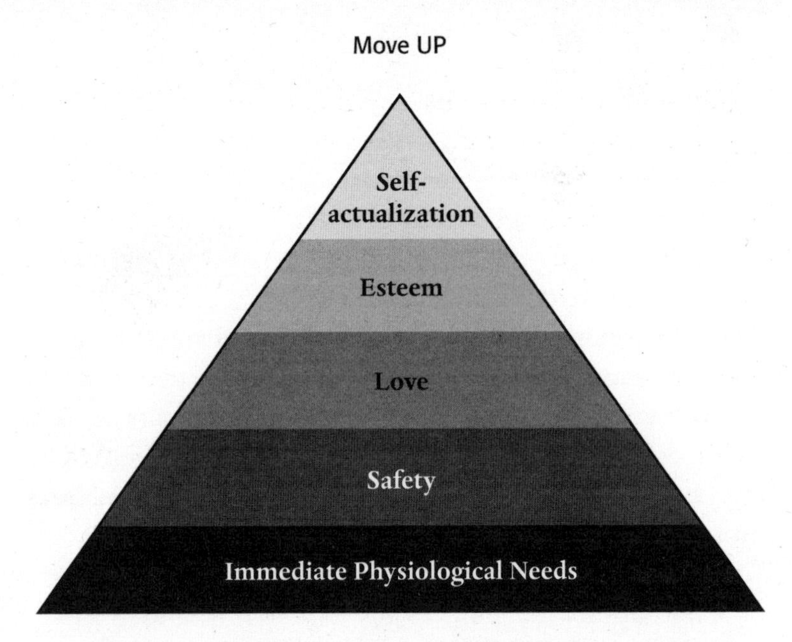

Maslow's 'Pyramid of Needs'

physiological needs are met, we then naturally move up and begin to worry about our safety, shelter and protection from threats. Once we meet our safety needs, we can then focus on what comes next in the hierarchy, our social motives. This includes our desire to have friends, co-workers, partners and a family who can show us affection, respect and admiration. At the top of the pyramid of motives we find our need for self-actualization, the desire to fulfil our creative potential.

Maslow's paradigm was avant-garde for his time, and several of his concepts have since been supported by sound evidence in neuroscience and behavioural economics. Maslow's guess was that people everywhere shared a set of universal motives, and he was right. He argued that the human brain does not operate according to one simple set of rules, but instead uses different sub-systems in order to accomplish different goals. There is now plenty of evidence that shows he was right there too.

But we have learned a lot since Maslow's day. Although his pyramid is worth analysing, it needs some updating in order to bring it into the twenty-first century.

Our pyramid shows the three levels of unconsciousness, distinguishing needs (that are biological) from wants. At the bottom we have our biological needs, basic needs we all share as human animals. Then we have our cultural unconscious (or the third unconscious), the level where we share things in common with others in a group within a cultural context. The cultural forces on this level answer to the biological needs at the base. Then at the top we have our individual unconscious. At this level, everyone is unique: these are our wants.

Why is our pyramid different from Maslow's? Because the

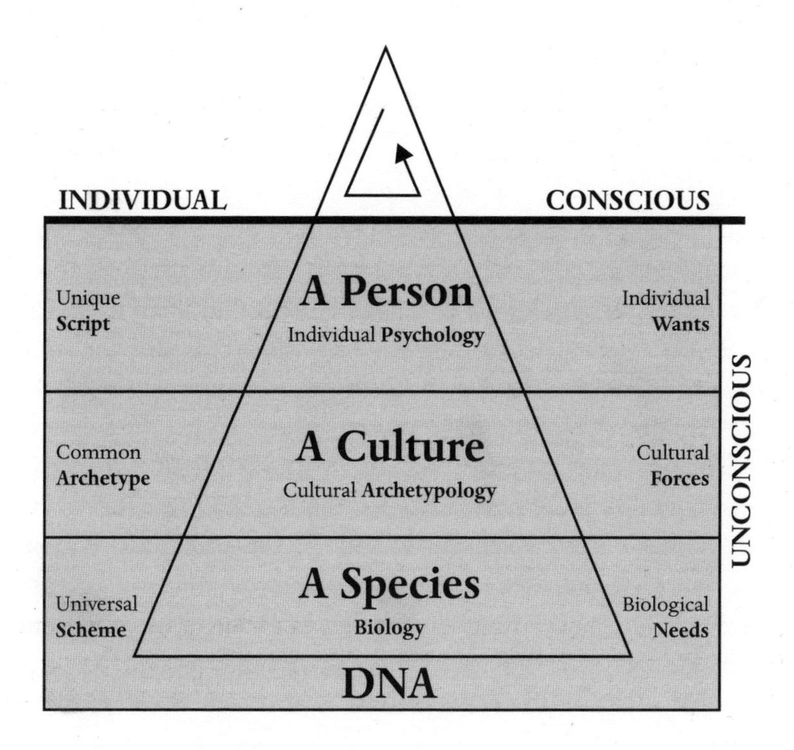

biggest problem with Maslow's pyramid is this: he did not understand the crucial importance of being bio-logical about human nature. He believed that the motives at the top of the hierarchy are somehow disconnected from our biology. There is now sound evidence that the highest realms of human creative genius are intimately connected to fundamental biological processes – processes that link us directly with the rest of the biological world. Understanding what motivates us, what makes us move, is after all a matter of life and death.

Over the last few years we have been conducting research into fundamental human motives. 'Motive' derives from the word 'move': what motivates us is what moves us. Our discoveries motivated us to recode Maslow's pyramid into a square with four axes rather than a pyramid: the bio-logical scheme of the Four S's.

$$\textbf{Bio-Logical} = \frac{\text{Survival} + \text{Sex} + \text{Security} + \text{Success}}{4}$$

The diagram opposite describes the Four S's of our biological being: survival, sex, security and success. Each S interconnects and interacts with the others in different ways. The arrows connecting one S to another represent the biological connection between the two: at the top of the arrow we describe the cultural aspects they share, and at the bottom the reptilian aspects they share. In the middle we find the compass containing the contradictions that can be found between the Four S's, contradictions that are deeply embedded within us and constantly at war with each other.

There are three important differences between our new paradigm and the old one. Firstly, Maslow's error was assuming that some motives take priority over others. So, instead of

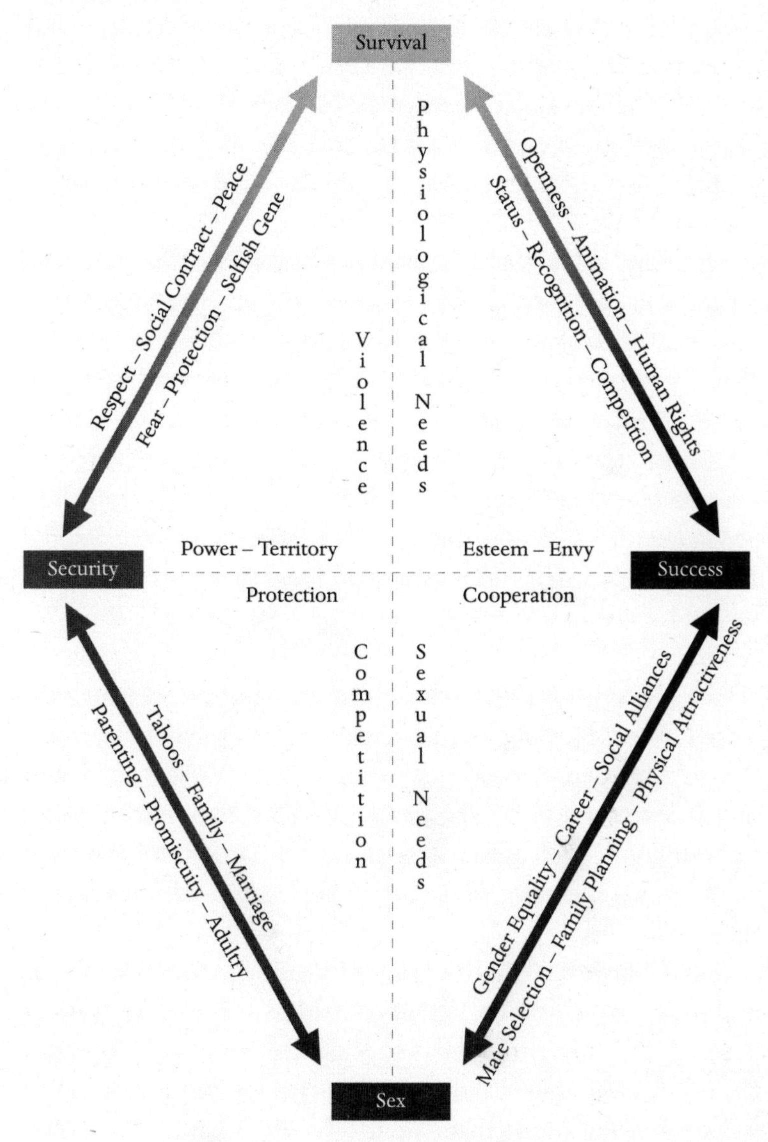

The Bio-Logical Scheme of the Four S's

stacking the crucial motives on top of one another, the goals in the new framework overlap, simulating a rugby scrum in 3D. In our new paradigm, motives do not replace one another, nor do they need to be accomplished sequentially; rather all the needs need to be met and are triggered by different situations at different times depending on the cultural and situational context.

Secondly, self-actualization is displaced from its hallowed position at the top of the pyramid. We are not saying that people do not experience all those 'higher' drives, but rather that, if we think in terms of our biological functions, a big part of what Maslow meant by self-actualization can be neatly folded into our 'success' category. Throughout history, people who perfected their creative performance or showed off their physical and/or intellectual capacities were often considered successful, and that often improved their chances of survival, sex and security. Ultimately, success is intricately tied to our biology.

Thirdly, Maslow did not understand the significance humans attach to sex. Maslow mentioned sex occasionally in his work, but he discussed it mostly as a simple physiological need, and had little to say about the other social repercussions of sex and reproduction, such as investing in the human capital of future generations, including the cost of health and education, and time lost.

What Maslow called physiological needs, such as hunger and thirst, are clearly designed to help us survive. As Maslow pointed out, there are lots of specific needs. Even hunger can be subdivided into particular nutrients at particular times when we need them most.

Our need for sex serves an obvious and essential survival function: without it we wouldn't be here in the first place.

Social, environmental and economic security is also an essential requirement for moving up. And success is the final bio-logical motivator and prerequisite for moving up: it is a drive closely linked to our biology.

So why do we need esteem and self-actualization? In our biological scheme, they translate into success, which translates into numerous benefits. Success means greater achievements, resources and fulfilment, which has obvious positive consequences for society and family.

Reproductive goals (sex) are the ultimate driving force behind much that moves us in human nature, such as being socially accepted and having a sense of belonging. Even attractive and well-connected personalities remain naturally sensitive to social acceptance and social rejection, and when they experience social isolation their emotional pain is registered by the same physiological mechanisms used to register physical pain. But in the end what moves us deeply is to be the heroes of our own stories.

All of these biological drives that move us aren't the whole picture. As we've said before, culture plays a role in all of this. How do our cultures deal with survival, sex, security and success? Let's find out.

8 *Survival*

SURVIVAL MOVING DOWN ——→ MOVING UP	
Human rights are not respected	Human rights are respected and promoted
No investment in arts and culture	Strong investment in arts and culture
Superstition rules	Knowledge and research rule
There is just one truth	Truth is a dialogue
Belief is more important than science	Science is more important than belief
The Golden Rule is: don't do to others what *you* don't want done to yourself	The Golden Rule is: don't do to others what *they* don't want done to themselves
Smoking is cool	Smoking is not cool
No awareness of our craving brain	Understanding and awareness of our craving brain
Culture of junk food	Culture of a healthy diet
Not concerned about health (i.e. diet, exercise)	Concerned about health and well-being

No promotion of cultural diversity	Promotion of cultural diversity
Physical, emotional and sexual abuse is tolerated	Physical, emotional and sexual abuse is punishable
Women, children and minorities are oppressed and limited	Women, children and minorities are integrated
Never question the status quo	Never stop questioning the status quo
Focus is on *what* to learn	Focus is on *how* to learn

We've talked about how the reptilian always wins, and how cultures can either embrace or reject this according to their value systems. But what about individuals? What about free will? If the reptilian always wins the battle within us, it's important for us to understand how it moves us, and how this in turn can help us move up. This is what being bio-logical is all about, and it's how we are going to determine the bio-logical value for the seventy-one countries in the R^2 Mobility Index. Survival is one of the variables that make up the Four S's of the bio-logical variable and in this chapter we are going to determine the values for survival.

From 1959 to 1969 it is estimated that between 16 and 17 million people (some estimates are as high as 45 million) died in China during the Great Leap Forward led by Mao Zedong. Millions more suffered and were persecuted and humiliated during the Cultural Revolution.[1] Survival was crushed during these hard times in China. Deng Xiaoping, having watched his family taking part in Mao's Cultural Revolution, described it as a disaster,[2] and so after Mao's death he worked to improve people's chances of survival.

Bringing Up Baby

A key reptilian notion of survival is that we cannot survive alone. If you take a baby at birth and leave it alone in the forest, the baby will die. Compared to other animal species, humans reach maturity very slowly. It takes years (even decades) until we can truly say we are self-sufficient. In a provocative sense, women don't finish the job.[3] Therefore we need an environment that nurtures us, a place where other people can look after us. And this is where family comes into play. You cannot survive without a family, without people

"Even when you were a baby you wouldn't eat."

feeding and protecting you – not just for a short period of time, but for many years.[4]

Since babies have no way of surviving on their own, they are at risk of suffering from anxiety. René Spitz, an Austrian-American psychiatrist, came across a case in an orphanage where the young babies were not eating properly and were losing weight rapidly. Spitz found out that, throughout the day, each baby had three nurses rather than one, so the babies didn't know how to connect with the nurses as they would with their own mother. Each nurse had a different way of interacting with babies, an inconsistency that confused the babies and inhibited them from being able to feed properly. Spitz found the solution: assigning a single nurse to care for each baby all the time. Continuity, consistency and constant care is essential for survival.[5]

You need a structure in order to survive, one that can provide you with food, shelter, protection and the conditions that may satisfy those basic needs on our chart. Those cultures that have strong family structures are more successful at helping us develop. For instance, cultures that are successful at preserving themselves even after a diaspora, such as the Chinese and Jewish cultures, have strong family structures: without them they'd have no way of transmitting their cultural values and perceptions.

The best family structures are those that have rituals, symbols, celebrations and a strong emotional investment. Rituals are usually reptilian: involving birth, death and marriage, or celebrations surrounding food like Thanksgiving and sabbath dinner. Every culture has its own rituals that celebrate the reptilian. Sometimes, the best function of these old traditions is their emotional bonding. But the key is to be aware that they are no longer necessary, that their function is purely social, and to use them for that purpose only, like the many young atheist Jews who still attend religious events.[6] We need rituals; they bring people together.

Superstition, however, is an enemy of education and human flourishing. When a culture attributes events to superstition, science and learning have little room for action. Watching the clock at 11:11 or being afraid of being on the thirteenth floor or in the thirteenth seat is pure superstition. We evolved to seek patterns, and that's why we remember 11:11 when seen on a clock and not 16:23 (there is nothing attractive about those numbers to remember). Out of all the times we look at the clock, we tend to remember 11:11 just because it is a pattern, four ones. And as far as the thirteenth floor is concerned, hotels are just fooling us. If spirits were hunting people on the thirteenth floor, I doubt they would stop just because the floor is labelled 'fourteenth' instead of 'thirteenth'.

These examples of superstition are silly, but when superstition is taken to an extreme, and cultures fail to prevent it, terrible things can happen that hamper survival. In 2008, four members of a family in India were accused of witchcraft and then attacked with stones and bricks. The police think they were buried alive.[7]

Superstition is relatively common in India, even among middle-income, educated families. They go to an astrologist for advice on when it is best to get married, and have cards or their palms read. It's worse for women. Some who are called witches get beaten and humiliated, by being forced to walk naked and to drink their own urine. Narendra Dabholkar, an anti-superstition activist, had since 1998 been promoting a Bill in the state of Maharashtra that would prohibit and penalize superstitious acts. The Bill was passed by the government in 2003, but it wasn't enacted until December 2013, after

Narendra had been shot dead in August of that year. Action is being taken in India, but sadly it has taken the life of many 'witches' and an important activist to spur it on.[8] India scores low in the Survival Index: it's ninth from the bottom. It takes more than a law in one state to change Indian culture. It is certainly one step in the right direction, although it would also help would they stop invoking gods with flowers and sandalwood paste before launching satellites.[9]

Other countries that are not doing to so well for Survival are Bangladesh, Pakistan, Sri Lanka and Kenya. In these countries poverty is a serious threat. In a place where the population can barely eat and access to drinkable water, not to mention health and education, is limited, survival gets more difficult.

Malala Yousafzai, the Pakistani activist who stood up for the right of women to be educated, was attacked by the Taliban. In the area of Pakistan where she lived, girls were banned from schools for a while, and many schools were destroyed by the Taliban. According to UNESCO, in 2010, the literacy rate of young Pakistani women aged 15–24 was 62.3 per cent, well below the average of 73.3 per cent for South Asia and the 78.4 per cent of the lower middle income countries category. In 2012, when she was on a school bus, a Taliban gunman shot her in the head. She survived and now lives in Britain, where she continues to advocate for girls' rights to an education. For a country where going to school can get you shot, the future in terms of Survival seems bleak.

Education is the first scenario that comes to mind when we talk about how to tackle socioeconomic problems, ignorance and religious extremism. Many argue that developing countries need more schools, better-paid teachers and smaller classrooms. The Programme for International Student

Assessment (PISA) run by the Organisation for Economic Co-operation and Development (OECD) offers some insights. Their tests put Shanghai in China at first place in maths with 612 points, followed by other Asian countries (other regions of China, Korea, Singapore and Japan) and some European countries (Liechtenstein, Switzerland, the Netherlands, Estonia, Finland at 519 points). The US does relatively poorly for being the largest economy, with 481 points. Does it really help to reduce the class size or pay teachers better, then? Apparently not.[10] The key is efficiency: teaching the teachers how to teach. Finland pays its teachers less than Germany or South Korea (although a little more than the US), and produces great academic results.

Survival Means Satisfying the Reptilian

After America and its allies won the war in Iraq and overthrew Saddam Hussein, the army's mission was not merely to defeat the enemy; they had to make sure that the Iraqis could survive afterwards. They needed to provide them with food, safety, infrastructure, jobs and so on. But soldiers are not trained for that: they were completely unprepared. The result was a complete mess. Before the war things weren't fantastic, but at least Iraqis could survive. If you can't provide for the basic needs of survival the reptilian demands, how are people supposed to trust you?

Switzerland is one of the strongest cultures in the world and it was created on the simple basis of survival. Eight hundred years ago, Switzerland was made up of different cantons with different dialects, but they had to unite and form a confederation in order to fight Austrian invaders. What they created was the Swiss culture, centred on the sheer need to survive. Today, Switzerland doesn't need an army, but every civilian is a soldier, ready for combat when necessary. They all have their uniforms and know how to use a weapon. The Swiss code is: 'We will never be invaded again.'

Russian culture used the natural environment in its fight for survival. When Napoleon wanted to invade Russia, he marched his huge army of half a million soldiers from Paris to Moscow. How did the Russians survive? They withdrew, taking all the food and wood with them. The French army didn't survive very long in these brutal conditions. Russians knew their environment and they used it to their advantage; they were familiar with their harsh winters. The Russian code for survival here was *not* to fight, but to let nature fight for you.[11] Let the reptilian win.

Pay Attention to the Rules

Cultures at the top of the index are very demanding when it comes to survival. Singapore, for example, is very demanding in terms of education: you will not receive a good grade unless you deserve it. Similarly in Japan, in order to read a newspaper you need to learn so much vocabulary that it is virtually impossible before the age of eighteen.

The best cultures for survival are those that are demanding. They may be perceived as oppressive and strict, but the rules are clear and people are expected to respect them. There is no survival without discipline. There's a good reason why these places are successful: they may not offer the kind of lifestyle you'd prefer, but they are good at helping people survive.

Survival has something to do with discipline, learning and being exposed to danger. It's the survival training that changes, not the species. It's the difference between a street cat and a house cat: both are cats, but their survival training is completely different.

In some African tribes, when a young boy turns twelve he is sent into the wilderness naked, painted white and only armed with a spear. Once all of his paint has washed off, he is able to return to the tribe. The point is that he has to learn how to survive on his own, away from the comfort of the tribe, and when he comes back he is no longer a boy but a man.

There are very few cultures that teach you *how* to learn. Many teach you *what* to learn, but not how to learn. Cultures that develop within you a love for learning, with an inquisitive, critical and even rebellious mind, are cultures that are going to help you move up. This is not at odds with discipline, rather part of it. You have to learn how to read the rules of the culture. At a Jesuit school, the rules can be so strict that you are even forbidden to speak, so when you make an error you are not told what you did wrong, you are just scolded. You have to learn how to read between the lines, read the rules of the culture. This is the kind of learning that will help you survive.

Each culture has its own code for learning. The French learn by criticizing (the *critique de texte*), the Chinese by copying (the tradition of the scribe), and the Americans by making mistakes ('you learn from your mistakes'). Americans rarely read instructions; instead they try something out, make a mistake, get upset (express an emotion) and then learn from it. As discussed earlier, emotions attached to experiences stay within our memory, which is a good way of learning. This is why you always get a second chance in America: what doesn't kill you makes you stronger.

You are rich because of all the things you have in your brain, not because of what you have in your pocket. What we want to convey in this book is that it is absolutely essential to first become aware, to absorb as much information from your cultural environment as possible: this knowledge is what will help you survive. If cultures are a survival kit, the survival kit must be decoded if it is to be useful.

If you feed a pigeon every time it pecks a button, it learns from the consistency of the routine. But if you feed a pigeon at random intervals of time, it quickly develops superstitious behaviour like hobbling on one leg or flapping its wing twice. The rules not only need to be clear, they need to be consistent.[12] But you will never learn all these lessons by staying in Kansas; you need to be adventurous enough to explore beyond your comfort zone, to take a trip to Oz. In France there is a saying that you can travel the world with just five cents; there is nothing stopping you. It's how you become an adult – by learning how to be on your own and to read the rules of other cultures in order to survive. Information is always available.

The Survival index

The table at the beginning of the chapter is a synthesis of all the qualitative elements behind the quantitative ones that we've found in cultures that either make you move down or move up. In order to translate these qualitative features into a quantitative reflection for the bio-logical component, we used health and education expenditure. These two variables show a high degree of correlation with social mobility. We also considered the 'Where to Be Born Index' developed by *The Economist*, to determine which quantitative elements allow individuals to survive within a culture. This last index considers crime, trust in public institutions, climate and wealth (among others). These are the variables that are most linked to subjective well-being.

So which countries are better at survival? The chart opposite depicts the Survival Index for seventy-one countries.

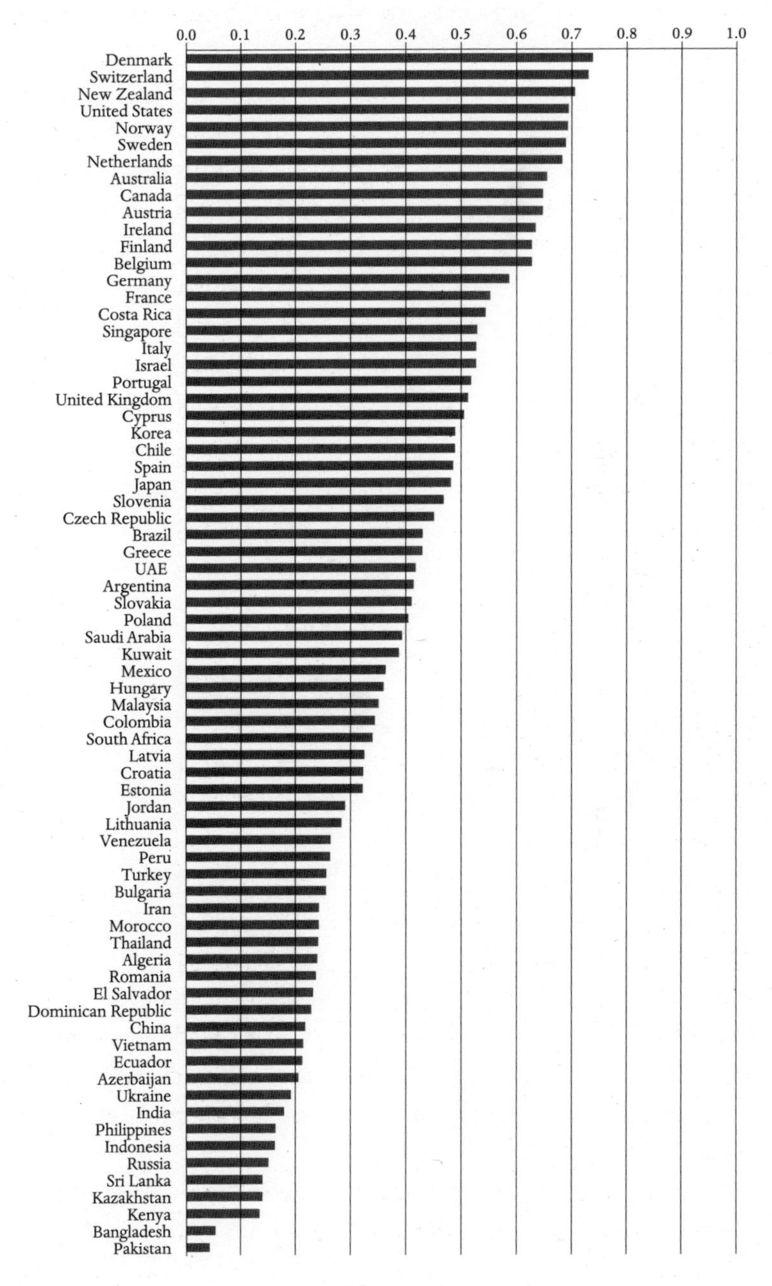

Survival Index

9 *Sex*

SEX MOVING DOWN ⟶ MOVING UP	
Pleasure is a sin	Pleasure is only pleasure
Taboos about sex	No taboos about sex
Virginity is sacred	Virginity is irrelevant
No sexual education at school	Sexual education at school
Women are not free to choose what to do with their bodies	Women are free to choose what to do with their bodies
Prohibit contraceptives	Promote contraceptives
Sexual life is strongly influenced by religion	Sexual life is about free, individual choice
Female genital mutilation is prevalent	Female genital mutilation is forbidden
Sex workers are treated disrespectfully	Sex workers are treated with respect
Society discriminates against individuals with different sexual preferences	Society is inclusive of individuals with different sexual preferences

Homosexuality is an issue	Homosexuality is not an issue
The concept of family is inflexible and closed	The concept of family is flexible and open
Men have greater power in family decision-making	Both parents have equal power in family decision-making
Arranged marriage	Marriage by choice
Many social barriers to divorce	No social barriers to divorce
Only one parent is in the labour force	Both parents can choose to participate in the labour force
Only men are encouraged to go to school and to study whatever they choose	Both genders are encouraged to go to school and to study whatever they choose
Career choices are based on gender	Career choices are not based on gender
Women have little freedom to choose the way they dress	Women can openly choose how they want to dress
Female promiscuity is frowned upon	Female promiscuity is perceived in the same way as male promiscuity

Adultery is severely punished and a public issue	Adultery is an issue only within the couple and is kept private
Women's rights are not respected	Women's rights are fully respected

Megan had a promising future with Richard, and she knew it. Her sex drive took the lead the night she cheated on Richard, and that ended up destroying that future. Would the outcome have been different had Richard been aware of her natural sexual impulses? Would it have been different had they both been French? What would have happened if they had been in Pakistan or Saudi Arabia and everyone had heard about her cheating on him? What would have happened if, instead of Megan cheating, Richard had cheated on Megan? How would different cultures react to this infidelity, and how is that relevant to moving up? In the end Megan's reptilian found a way out: it expressed itself through the one-night stand with Bradley.

If we want to move up, we need to be reptilian and understand sex from two perspectives: on the one hand, our ability to have sex for recreational as well as reproductive purposes; and, on the other, equality between the sexes. If we look into the reptilian dimension, the game between men and women is about sex. No sex means no children, as simple as that. And no children means that nothing will be passed on to the next generation. No sex means not being bio-logical. The main reason we have sex is precisely this: it is the most fundamental means for a species to survive.

What does the reptilian have to say about sex? Around the

world, inside the reptilian brain, women want to be chosen, and men don't want to be rejected. That's the code. The way women will present themselves is that they want to be chosen, they are afraid of not being chosen; whereas men are afraid of being rejected. The prom ritual in America is a perfect example of this: the girl waits around for a guy to ask her out, and the guy is too shy to ask his dream girl, afraid she will turn him down. Some cultures even go as far as promoting their daughters to society, with coming-of-age parties and *quinceañeras*. Cinderella waited around for Prince Charming to arrive and choose her rather than her stepsisters. So women have a tendency to wait. This is their biology. The egg waits while the sperm moves. Some cultures are going to try to change this.

When it comes to getting down to sex, how do the reptilian, limbic and cortex respond? In an experiment, a group of young people were left in a cave without any light for three days. They quickly paired off and had sex the whole time. Their reptilian dominated – it was about the now, not caring, not judging, just getting on with it. Later, they ran the same experiment with a second group, giving them light on the second day, and found that only a few paired up. For a third group, they had the lights on the entire time, and the participants had met on a regular basis for six months preceeding the experiment. No one had sex. That was the cortex at work, thinking in the long term – giving them time to think and get to know the other people and judge if they wanted to engage in anything with them or not.

Our reptilian brain demands that we think about sex.[1] It helps us search for the appropriate mate in order to increase our chances of survival and success, as well as those of our offspring. Men and women differ greatly when it comes to mate selection. Evolutionary psychology has shown us that men tend to look for young, attractive and fertile women, and women tend to look for strong, older and resourceful men.[2]

The emotional bonding attached to sex also serves as a survival function, whereby a couple that has a greater emotional bond is more likely to stay together and raise a stable family. With so many survival benefits, it should come as no surprise that, beyond the act of procreation, sex is also a pleasurable and fun activity. This is why, when a couple is trying to get pregnant and their doctor recommends having sex at certain times of the day in certain positions, sex becomes less fun and can even become painful. One loses the pleasure of losing control. Sex is more than sex and reproduction. It is bonding. It is acceptance and being accepted. Sex is complex.[3]

When we talk about reproduction, sex can mean having a child, family planning and also recreational sex. In all cases, it involves establishing relationships with other human beings. But even though sex is one of our most basic instincts, it can also be dangerous, inhibiting our growth or movement. From unwanted pregnancy to sexually transmitted infections, sex is risky.

Although we are programmed to be baby-making machines, and to actually enjoy sex, our biology can sometimes inhibit us from moving up if we are not logical about it. When individuals start families at a young age, either voluntarily or by accident, they can limit their opportunities for moving up. The reality show *Teen Mom* follows young women across America who decide to keep their children, often struggling to finish high school, get a job, establish a healthy relationship and give their babies a stable environment.

What makes some people more successful at reproducing their genes? The answer is a combination of biology and culture. Culture is our biological strategy, based on cooperation, empathy, the sense of belonging, shared information and interlocking roles. Cultures were made to be a new and better way of relating to each other. Successful cultures are those that are constantly moving between the tension that separates life and death. Death is down, and life is up. There is a perpetual tension between life and death, between Eros and Thanatos. And most of the time this has worked.

Jared Diamnod imagines in his book *Why Is Sex Fun?* how a dog might perceive human sex:

> Those disgusting humans have sex any day of the month! Barbara proposes sex even when she knows perfectly well she isn't fertile – like just after her period. John is eager for sex all the time, without caring whether his efforts could result in a baby or not. But if you want to hear something really gross – Barbara and John kept on having sex while she was pregnant! That's as bad as all the times when John's parents come for visit, and I can hear them too having sex, although John's mother went through this thing they call menopause years ago. Now she can't have babies anymore, but she still wants sex, and John's father obliges her. What a waste of effort! Here's the weirdest thing of all: Barbara and John, and John's parents, close the bedroom door and have sex in private, instead of doing it in front of friends like any self-respecting dog.[4]

Sex is often a taboo in many countries of Latin America, the Middle East and in the US. Whereas in France and Germany discussing sex with your teenage kids might be awkward but still considered normal, many families in the US would be offended if such a topic come up at the table.

The reptilian commands us to have sex: that's why it's pleasurable. This is true whether you are Argentinean, American, Iranian or Portuguese. It is in our biology, and it's up to cultures to decide how they tame or express this desire. They can acknowledge it or they can pretend it isn't there. In the end, the reptilian will win, and will get expressed.

The reptilian is also expressed through the Internet and smartphone apps. Since prostitution is illegal in most countries, the Internet provides a way to buy sex and satisfy the reptilian without getting exposed to arrest on the street. Pornography websites are even solving some of the market problems: clients can read reviews of the quality of the service, sex workers will know better which clients to avoid, and pimps are disappearing. In countries where prostitution is legal, such as Germany – which is in the top ten of our ranking – both prostitutes and clients can go to the police if they believe someone is being abused. Contrast this with countries which will punish the client, the sex worker or both for going to the police.[5]

In Iran, couples are finding the gaps they can use to have relationships without having to get married: since the government can't control what happens inside the home, they can't prevent people from having sex. Iran's parliamentary research department reported that 80 per cent of unmarried women had boyfriends. What some advocate therefore is the adoption of the ancient Shia practice of *sigheh*, or temporary marriage.[6]

Genderally Speaking

> Sure he [Fred Astaire] was great, but don't forget Ginger Rogers did everything he did backwards . . . and in high heels!
>
> Bob Thaves[7]

We usually say, foolishly, that men and women are equal, but let's face it: biologically, men and women are not equal. We are different. We may be complementary, but we are not the same. Of course we should have equal rights, equal opportunities and so on, but it's a mistake to say that women and men are the same, because they are not.[8]

Working with female executives around the world, we have found a common denominator among these high-flying professionals: they feel not only that they have to behave like men, but they have to be stronger than men too. There is no room for weakness. There are behaviours that are taboo for them, such as being moody or emotional. This completely goes against women's inherent cyclical lives. To be normal in the business environment, you have to be abnormal as a woman. And that is one of the crises our business world faces today: a macho cultural structure that doesn't allow a woman to be a woman.[9]

We should respect the reptilian, but we have different reptilians. As a man, you may not always understand your wife, and you don't need to: you just need to love her, and vice versa. The most important thing to remember when talking about relationships between the sexes is the role of the reptilian – in feeling attraction and sexual desire, and ensuring the survival of our species through reproduction. The reptilian level is the most important, then the limbic level (love and delay), and finally the cortex level (understanding, long-term planning and so on). The big mistake happens when men and women try to understand each other, label each other, and control each other.

Some cultures don't try to understand, but others do. American culture is a very reptilian culture, and a very male-centred culture. They may be able to go to the moon, but they sure don't understand women, and, even when they say they try, they don't. Women will try to get men to understand them and get in touch with their emotions, but American men just want to watch their Sunday night football with a beer in one hand and the remote control in the other. It's a culture that is not favourable for women, and a culture that makes the connection between the sexes very difficult.

The question of gender equality has only become central in recent decades. The discourse surrounding equality amongst men and women has been fundamental in understanding the sexes better as a society. It is obvious that there are biological differences, but they are more quantitative than qualitative. When it comes to gender, it makes sense to include gender equality in the mobility equation. Ignoring the rights of either gender translates into a loss of talent and skills. As we've mentioned before, biology plays a big part in our differences, and especially in what distinguishes men from women. But equality of opportunity and choice are crucial.

In the Sex component of the index, not surprisingly, the Nordic countries head the list, led by Sweden. The Nordic countries in general, and Sweden in particular, have had very inclusive policies towards women. Forty years ago, the Swedish government began a policy of granting parental leave. Today, parents get 480 days of paid leave, out of which at least 60 must be used by the father and 60 by the mother. In 2012, men took 24 per cent of parental leave. (Parental leave also applies with adopted children.) Not only does this policy contribute greatly to closing the gender gap, but it also promotes women's careers. A very renowned doctor and researcher once told us over dinner everything he does: he manages to do research, deliver classes, practise as a neurosurgeon (while teaching students) and play with his kids. When we asked how he managed to do all these things, his wife swiftly replied, 'Well, he has a wife.' Parental leave helps to level the playing field for women's careers. If it's the father who stays at home for a while, the mother can also work without having to choose between career and family.

Wasting Half the Glass

We need an equal proportion of men and women in order to survive, which is why it makes no sense to limit half the population. Biology cannot be the justification for a culture where women do not have an equal opportunity to make a choice. A culture that doesn't take advantage of its women is a culture that kills mobility.[10]

Qatar, a Muslim country, is actually doing a lot to improve the integration of women in society. There has been a huge shift towards providing their women with quality education. Most university students are women (which is not good either – what are the men doing?); this says a lot especially when one compares Qatar to its neighbours, who are still failing to integrate their women into society. Take Sheikha Mozah, a living example of Qatar's radical change, who is the head of Education City, and the only royal to have a nine-to-five job.

During the 2012 Olympic Games in London, the International Olympic Committee was about to ban Saudi Arabia from participating in the Games because its team had no female athletes. At the last minute, Saudi Arabia sent two women to participate – but neither had ever lived in the country; they had both been raised abroad. In Saudi Arabia women are strictly forbidden to practise any official sport in public schools or to take part in national games. This restriction is not literally based on the Koran, but rather on a particular cultural interpretation of it.

Women can't vote in Saudi Arabia, and can even be sentenced by the courts if they are sexually attacked. A nineteen-year-old girl was raped by seven men in the city of Qatif. She was sentenced to ninety lashes for being in a car with a man who was not her husband.[11] The judges' sentences are completely discretionary, so they vary from judge to judge. What would have happened to Megan had she been Saudi?

Manal Al-Sharif is a Saudi woman who dared to drive in a country where women are not allowed to drive. She posted a video of herself driving in Saudi Arabia, and as a result got arrested. She was allowed out on bail only after promising never to drive again. The driving ban makes it difficult for women to live independent lives. Women need the permission of a male guardian to travel or get medical treatment. Saudi Arabia is at the bottom of the Sex ranking, and it scores poorly in almost any other gender index.[12]

If you were to extrapolate this discrimination against women and apply it across the globe, the world would suffer a terrible loss of human capital. Women make up more than half the world's population, a potential waste not worth contemplating, solely based on archaic male judgements about female competence.

We repeat: half the population is female. Failing to provide for every member of society is a waste of talent and opportunities. Discrimination towards people within our own species in general is backward, setting aside the question of human rights. Gender equality is a necessary condition if we are to enhance mobility. Policies such as creating equal employment opportunities, increasing the participation of women in politics and high-level positions, enfranchising women, giving equal maternity and paternity benefits, encouraging men into the fashion and gastronomy industries – all reflect a culture's disposition to increase people's opportunities for moving up. How are we going to do that if we are limiting success to half the population?

India, the third country from the bottom of the Sex index, faces serious problems with gender. In 2012, a brutal case of rape in the city of Delhi in India provoked outrage and public protests. Crimes against women are under-reported in India because women don't trust the justice system, and also know that their reputation is at stake. They are often blamed – accused of having provoked the aggressor – or disbelieved. Nevertheless, according to the National Crime Records Bureau of India, there were 33,707 cases of rape reported in 2013 alone.

India has taken action and is trying to empower women. In 1993 a change in the constitution mandated that one third of *pradhans*, the leader of a village council, with substantial power in influencing decisions, should be women. Since the allocation of this one third was randomized, an evaluation could be made. It was found that women leaders better reflected their gender's preferences when policy decision had to be made. In communities in West Bengal, women were more concerned with drinkingwater and roads than other problems. The villages that had women as leaders built more water facilities and improved the roads more than others led by men.[13] This is evidence to refute those who think that men can effectively represent the interests of all women.

The Heels and the Beard of Cultures

Every culture has a masculine and a feminine side, and the relationship between the two must be understood. Femininity is about integration, about creating life by integrating something within your body. Masculinity is about separation: creating life by putting something outside your body. Women define things by inclusion and communication, using many more words to describe things than men, and taking account of a variety of considerations when making decisions. Men are linear, whereas women consider seemingly extraneous details, making them matter.

Some cultures are very feminine. Brazilian culture is quite

feminine, since it's very much into integration. Brazil is racially integrated: every individual is a mix of different races – native indigenous, Spanish or Portuguese (which are themselves a melting pot of races) and African. In short, they are hybrids, which is also biology's bid for a better chance in the game of survival. Even their food reflects this trend: the national dish, *feijoida*, is a mix of all kinds of ingredients; if there are extra guests at the table you just add water and there's enough for everyone.

Around the world, women are cleaner than men because women are in charge of creating life; they have very special equipment to take care of which means that they are high maintenance. Since Brazilians have a more feminine culture, they are cleaner. They say, 'We might be poor, but we are clean.' Clean comes up again, and this is not only reptilian, but also a feminine dimension. Women have to make sure that everything is in order and take no risks for the children they raise. *Mulher guerreira*, or 'woman warrior', is a woman who fights every day to make sure everything is clean, the children are fed, there's a roof over their head; she has the final say. The unconscious female reptilian brain is about maintenance, cleanliness and odours. Even if conditions are very difficult, they are always clean, and Brazilians have assimilated that behaviour in their culture.

If we compare Brazil and Argentina, something very interesting comes up: differences in dance. Argentina has tango, a very sensual but nevertheless male-dominated dance. The woman goes around the man seducing him, but he always has the power to attract or reject her. Brazilian samba, on the other hand, is much more female-centred: the men dance around the women. These two Latin American cultures share many similarities, but they are also vastly different.

Cultures that enjoy women are cultures that take pleasure in the reptilian, but they are going beyond and engaging the cortex. Parisian women are empowered women. They are fashion-oriented, and famous for having a talent for pleasure. Parisian women are taught from a young age that it doesn't matter what you have, it matters what you do with it. A Parisienne can just grab a few things from her wardrobe and put them together beautifully. She doesn't need to exceed her credit card limit to feel fabulous.

Every time I hear Hillary Clinton speak,
I involuntarily cross my legs.

Tucker Carlson[14]

Let's talk about Hillary Clinton, and her wardrobe. Why does she never wear a skirt? Because she cannot succeed as a woman in the male-dominated world of politics. America has a hard time with the intelligent woman. A bimbo, on the other hand, is beautiful and dumb, the perfect combination. But American men are very afraid of intelligent women, and that's one of the handicaps for Hillary because she is too bright and not feminine enough. But on the other hand, America would also be uncomfortable if she was *too* feminine.

If we look to Japan, there is a stereotype that doesn't correspond to reality: that Japanese women are submissive. We've been doing a lot of work in Japan, and our experience is that Japanese women are in charge of everything reptilian. The men go to work, it takes them two hours to get to work on the subway, and when they get home they are tired – they eat a bowl of soup and go straight to bed. Japanese women are in charge: they control the money, the house, investments, everything. Locally, women are sheriffs, they do construction work, they go to university and get an education. This is not the clichéd image of a submissive geisha. But part of the culture is a trade-off: once a Japanese woman marries, she cannot work outside the home, because that is her domain and she is going to be solely responsible for it.

The hard data that captures the characteristics presented on the table at the beginning of the chapter is measured through counting, for example, women's participation in employment. More equal societies will have a higher proportion of women in the workforce. Even in powerful positions such as politics, women should be present not only for the sake of democracy, but also because no one should be excluded from such activities based on gender rather than merit or performance.

Family planning and mothers' health are important dimensions of sexual and reproductive health, and are also accounted for in the Sex index. We don't have to elaborate a new index in this matter, since the UN elaborated the Gender Inequality Index, which reflects all these features. Countries are ranked according to this index in the table on p. 142.

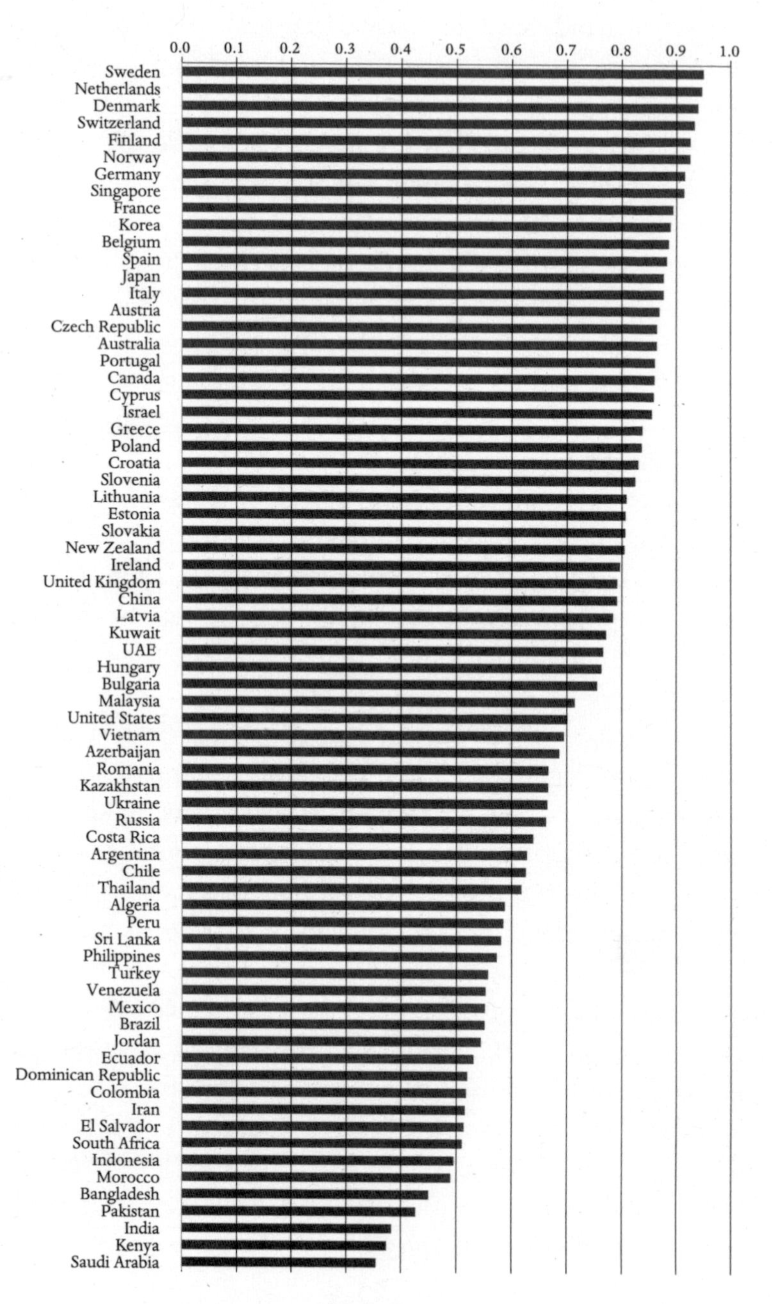
0.0
0.1
0.2
0.3
0.4
0.5
0.6
0.7
0.8
0.9
1.0
Sweden
Netherlands
Denmark
Switzerland
Finland
Norway
Germany
Singapore
France
Korea
Belgium
Spain
Japan
Italy
Austria
Czech Republic
Australia
Portugal
Canada
Cyprus
Israel
Greece
Poland
Croatia
Slovenia
Lithuania
Estonia
Slovakia
New Zealand
Ireland
United Kingdom
China
Latvia
Kuwait
UAE
Hungary
Bulgaria
Malaysia
United States
Vietnam
Azerbaijan
Romania
Kazakhstan
Ukraine
Russia
Costa Rica
Argentina
Chile
Thailand
Algeria
Peru
Sri Lanka
Philippines
Turkey
Venezuela
Mexico
Brazil
Jordan
Ecuador
Dominican Republic
Colombia
Iran
El Salvador
South Africa
Indonesia
Morocco
Bangladesh
Pakistan
India
Kenya
Saudi Arabia
Sex Index

10 *Security*

'There's no place like home, there's no place like home, there's no place like home.'

Dorothy in *The Wizard of Oz* (1939)

SECURITY	
MOVING DOWN ⟶	MOVING UP
Legalize prohibition	Legalize freedom
Culture of disrespect and distrust	Culture of respect and trust
Poor community strength when facing a common threat	High community strength when facing a common threat
Justice by one's own hand	Justice is institutionalized
Police corruption is the norm	Police corruption is not tolerated
The population is at high risk from natural disasters	The community works together during natural disasters
Environment of risk aversion	Environment of risk-taking

Being a politician is perceived as something corrupt	Being a politician is perceived as something respected
The government controls the market	The government allows a free market
People don't trust institutions	People trust institutions
Citizens don't have access to public information or transparency	Citizens have access to public information and transparency
The government doesn't care about the human rights of their citizens	The government respects all the human rights of their citizens
Prohibition of recreational drugs	Legalization and regulation of recreational drugs
Gangs are seen as a way to success	Gangs are seen as something bad for society
The police intimidate people	The police are respected by people
Expropriation is common	Expropriation is unacceptable
Contracts are breached	Contracts are adhered to
Property rights are not well defined	Property rights are well defined
There is no punishment for kidnappers	Kidnappers face justice

Impunity is the norm	Impunity is not tolerated
Domestic violence is tolerated	Domestic violence is punished
There is no opportunity to be economically independent	People have the opportunity to be economically independent
Social security does not reflect tax revenue	Taxes are effectively and efficiently used for social security

Security is a basic requirement for survival, and thus it's a prerequisite for moving up. Through our studies, we have found that safety is feminine and security is masculine. When a company says they have a department of safety and security, they are mixing two completely different concepts.

China has grown a lot since the times of Mao, and China's recent development is thanks to Deng Xiaoping and his spirit of reform. China has a long way to go. It is now in the middle of the Security index, and it would soar higher were it not for frequent expropriations (particularly in rural areas) and the vagueness of the law regarding property rights.

When the Beijing Olympic Games were about to take place, many people lost their homes to make room for a more modern infrastructure. Those people received insufficient compensation and the government failed to relocate them as promised. Two women in their seventies were applying for a permit to protest, in the area of Beijing designated for protests, about the homes they had lost during the Olympic Games when they were arrested and sentenced to being 're-educated through labour'. This is only one of many cases in China, and the Chinese still have to work on refining their laws to give judicial certainty and respect for human rights.[1]

Giving Up the Ghost

There's no place like home. If you're American you know that home is the best place in the world and it is safe. When you are inside, when you are at home, have some hot food on the table and the doors are closed, you are safe. Safety relates to the female: it's a feminine concept depicting motherhood and the nurture and care of the mother. Mothers feed you, care for you and create an environment where you can feel protected. You can let go, let your guard down, leave your gun at the door and relax.

You feel safe when you no longer have to be in control, and that's a feeling a mother gives. The mother figure is supposed

"Hello, Dad? You were right about the world. I want to come home."

to care for you no matter what; she's not going to judge you. We can apply this concept of safety when we talk about hospitality. Hotels, like homes, are supposed to make you feel safe. For example, the waitress isn't going to poison your drink, your room is private and only yours while you're there. Scent is also attached to the feeling of home, and restaurants that can pull this off tend to be more successful. When you walk into the cinema, the smell of popcorn immediately makes you want to sit down in a big comfy chair and settle in to watch a movie.

Cultures vary when it comes to the safety of the home and its cosiness. The British are famous for their cosy homes, the small cottage with a fire going, a comfortable sofa. From Hogwarts in *Harry Potter* to the Shire in *Lord of the Rings*, we associate British homes with cosiness and safety.

In American culture, the concept of 'home' is based around the kitchen. In fancy American homes, the kitchen is the largest space, with an island in the middle and benches around for people to gather while Mother cooks. It's about food, about feeling full. The ritual of Thanksgiving provides that safety because you have good food and you are surrounded by people you know and care about. Southern home cooking is famous for its warm, filling and hearty dishes, the kind your grandmother might have made. The American brands Betty Crocker, Sarah Lee and Aunt Jemima are all archetypes of the mother in the kitchen cooking up something delicious. Every American woman wants to be her; she is the queen of the kitchen. She feeds her family, and makes them feel safe and satisfied.

At home, the door is a big safety symbol. Through working on doors, we've found that a door should speak to you. If you close the door and it doesn't make a noise, you assume that it hasn't closed properly and you are not safe yet. Studies have shown that the perfect sound a door should make is the sound of three clicks: three clicks for safety. The more clicks, the safer you are: you don't even need to check if the door closed properly, it's already told you it's closed.

When it comes to the means of entering or leaving your home, cultures vary greatly. France has been invaded multiple times by its neighbours, and so the priority is to have high walls, small windows and a strong door so no one can come in. In American culture, the number one fear is of getting stuck inside if there's a fire. American homes are not usually made of stone, so they tend to catch fire easily, and the priority is to be able to escape quickly. Americans have large doors and windows, and even doors with windows that they can peer out of. There are fire escapes leading up the side of apartment buildings, which make perfect access points for a burglar – but the priority is being able to flee if necessary. In California, it is forbidden to put bars on the windows in case a child gets locked in during a fire and the firemen need to break the windows to rescue people. The reptilian priority for Americans is keeping your family safe and allowing them to get out rather than stopping a burglar from coming in and stealing your valuables.

When American tourists visit Mexico City they are amazed by the high walls that surround the homes. Apparently, residents of this city feel more secure if they have a brick wall around their homes to keep out thieves, strangers and maybe even some friends. But Americans would probably feel trapped in such a building, and they wouldn't know whether they were protected or just isolated from the rest of the world. It seems normal for Mexicans because these walls have been a common sight throughout their lives.[2]

Stop Chasing Shadows

If we look instead at security, we are talking about a completely different reference system. We need to be clear that when we talk about security, we are talking about maintaining national borders, and protecting ourselves from harm, attack and any crimes that could destabilize daily life. This is masculine. When we talk about 'homeland security', we are talking about making sure enemies don't come in. On a reptilian level, when a man and a woman have a family and a home, it's the woman's duty to make sure the home is childproof, while it's the man's duty to make sure it's secure from potential invaders. Women are in charge of making their place safe, and men are in charge of making sure no one can enter their safe place.

Security means weapons, guns and castles; anything and everything that can keep people out of the safe place. The whole idea behind the Great Wall of China was to stop the Mongolians from invading: that's security. When you go to a nightclub, the bouncers have shirts that say 'Security', not 'Safety', because they are preventing unwanted individuals from entering the safe space.

The whole point of security is accumulating enough backup in order *not* to fight. In *The Karate Kid*, Mr Miyagi teaches Daniel how to fight so he wouldn't *need* to. It's about accumulating enough weapons, soldiers, bombs and so forth, so that the enemy is too scared to attack you. It's the quantity archetype of masculinity – the reptilian again.

Security can be absurd, though, too. We've arrived at the point where so many nations have amassed such an incredible amount of weaponry that, as we've seen with the Cold War, it just ends in stalemate. This is good in the sense that it avoids death on a large scale, but not all the nuclear weapons in the world could have stopped the terrorists from attacking America on 9/11. It takes just a few extremists, with not many resources, to completely throw a country off its notion of what security really means. Barry Buzan, a highly regarded authority on security and securitization, says, 'States generally, and some governments in particular, need threats in order to justify their existence.'[3]

For the purpose of this book, it is interesting to look at the cultures that have arisen around the concept of security; around the threat of danger from the outside. For Israel, creating a state in 1948 was extremely difficult. It's a country that has never stopped fighting since the very beginning. People have no choice, and are ready to fight, so despite any differences amongst each other they are constantly prepared to fight together in order to survive. Things would be very different if Israel suddenly no longer had any threats to contend with.

Security today has changed drastically. If we look at the statistics, the world has never been so secure. However, the media hypes up people's perceptions: we are in an age of fear, and people feel less safe now than they did two hundred years ago.[4] The reality is that fear is a political agenda. This is because nothing unites a society more than a common enemy. Hitler used the Jews, nineteenth-century Germany used the French and the Russians. Israel is still so united because they are surrounded by enemies. They operate on reptilian priorities.

Why do we like what we like? As our friend Eduard Punset said, 'The answer to that question is simple, albeit unexpected: we like everything that makes us secure',[5] that is, the people and institutions in which we trust. It's not an accident that 'peace' and 'security' have the same Proto-Indo-European root, and what lies behind these words is the idea of trust. Without trust, there is no peace and security, and vice versa.

Security

Humans need to feel safe. Emotional security is crucial. Imagine being a Jew in Nazi territory and living with endless fear every day. Imagine walking to work every day and having to watch your back for fear of being robbed. Imagine being unable to go out and relax at a bar or restaurant because of the threat of violence. The Indo-European root of the word 'angst' comes from the word '*angu*', which means restriction. We feel restricted from doing things. Anxiety is a handicap that prevents freedom of thought and movement.

The US government recognized this necessity after 9/11. Security in airports is crazy now. You have to take off your shoes and belt, and then pass through a metal detector, sometimes several times. Although a costly policy, this security mania does make people feel safe. The security expert Bruce Schneier doubts how effective this method really is for catching terrorists, but it certainly makes us feel in control of the situation and therefore more safe within our homeland.[6]

Fear of terrorism, robbery or getting killed is one dimension, but fear of not having a job, of not having health care, of losing your home when a hurricane strikes, or fear of having your house or business expropriated are also important elements of security. When there is a personal crisis in your life – you lose your job or fall ill – the fear of not getting by is very high. This is where social security is important. Nordic countries are actually very good at it. Sure, their governments charge an incredibly high tax rate relative to income, but you can be certain that you'll be covered in the event of a crisis, and you can feel more or less at ease in the knowledge that you'll be looked after.

Take Sweden as an example: a high tax rate allows the government to spend vigorously on high standards of living. Sweden is an interesting case because there is a lot of protection: the state does a really great job of protecting everybody with a good healthcare system, a good education system, pensions and public transport. Women have a lot of freedom because the state takes care of their children. Although they enjoy gender equality, contrary to the cliché they are not very reptilian in the sense that they are not very affectionate: some would say that it's almost like looking at a nice dish that has no flavour. They are beautiful people who aren't known for their warm nature. In the environmental sphere, only 1 per cent of Swedish trash ends up in rubbish dumps. The rest is burned to provide electricity for cities.

We cannot say the same for some Latin American and Asian countries, whose governments are not prepared to act when natural disasters take place. In 2010 an earthquake in Haiti practically destroyed the country, devastating the already poor population. As with the tsunami in South Asia in 2004, international aid came quickly, but these countries (and their cultures) were not prepared for such disasters.

Now imagine you live in Venezuela, a country with one of the highest crime rates in the world (as high as 53.7 murders per 100,000 inhabitants in 2012, according to the United Nations), and with a government crowding out investors not only from abroad, but also from within its borders. Rules are not well defined. The government can change its mind at will. Entrepreneurs find it hard to innovate since they have no way of knowing who owns what.

Respect for property rights is crucial if a country is to move up. With such rights properly and fairly established (and with law enforcement to maintain them), the population can free up their inventive potential and put their resources to work. How can you create something new if you don't really know what belongs to you? That is Latin America's dilemma. In a region where so many ideologies cross paths, nothing is clear, and therefore they remain stagnant. Argentina took a step backwards when it decided to seize control of the oil company YPF, which had been founded by the Argentine government and then privatized in 1990. It's difficult to believe that property is in permanent danger in a country that was once richer than Germany and Italy, where everybody wanted to migrate to.

Bangladesh, second from the bottom in the Security index, has a reputation for corruption. It's 144th in the 2012 Corruption Perception Index prepared by Transparency International. Niko, a Canadian-owned energy company, was first rejected when it applied to operate in Bangladesh for not having the expertise and financial resources. It later got a contract with a state-owned company, began drilling for gas and caused an explosion. Trying to control this first accident, they caused another explosion. Niko refused at first to pay for damages (they eventually paid), and later gave an expensive car as a gift to their partner company, which was then given to the energy minister, who was responsible for overseeing the issue of compensation.[7] Bangladesh made it onto the world's media when a clothing factory collapsed killing 1,129 people and injuring 2,500 more: you can imagine the feeling of insecurity among garment factory workers. This tragedy was followed by new safety laws, to try and prevent such a thing happening again.

Private property, combined with competition and the legalization of freedom is a great start towards a healthy market.[8] The point is to have rules and respect them: that is property security. One of the benefits of private ownership is competition. When two individuals produce the same goods, they will compete, resulting in lower prices and better quality: happy consumers, happy entrepreneurs.

China is currently attracting a lot of capital. Big companies manufacture their products there, not only because it's cheaper, but also because they are finding a huge market in the most populated country on Earth. Part of China's great success is due to the adoption of foreign technologies. After the Cultural Revolution, China was left with old, inefficient technology. Under Deng's policies, the Chinese copied technologies instead of creating them (which makes sense, of course), which explains part of its rapid growth. They also began to copy products, which threatened property rights. Some of the cheap items found everywhere in the world (for example some smartphones, tablets and televisions) are imitations of other products whose creators invested in their innovation. As Jack Perkowski claims in *Forbes*, some companies can't afford not to go into China.[9] Not only for the cheap labour and large market, but also because if they don't, the Chinese can figure out a way of copying their product and selling it not only within China, but also abroad, reducing the market in China and beyond for the company that first produced the item.

The United States has strong protection for property rights, but it scores surprisingly low in Security. Countries such as Costa Rica, South Korea, Mexico and Chile score higher. The United States is the biggest economy in the world, so why the low score? The level of inequality in the US (the highest amongst the developed countries) doesn't help. Americans don't save money: they consume and live in debt most of their lives. Student loans are high, and take a long time to pay back. In other countries, even those with fewer opportunities for higher education than the US, culture dictates that, if you can send your kids to college, you will. Instead of having moving up in mind, in the US the decision is more about 'Is it worth the loan?'

Another feature that hampers Security in the US is its criminal justice system. The US has the highest incarceration rate in the world: by 2012, it was 707 per 100,000 inhabitats.[10] It is a racial issue – prisons are largely populated by black Americans. If you are black, you are far more likely to get arrested for marijuana possession, for example, than if you are white. Iowa is the state with the largest black–white arrest disparity; blacks are 8.34 times more likely than whites to get arrested for marijuana possession.[11] A report prepared by The Sentencing Project in 2013 says, 'Racial minorities are more likely than white Americans to be arrested; once arrested, they are more likely to be convicted; and once convicted, they are more likely to face stiff sentences.' It's worrying that in the richest country in the world, black males remain six times more likely to be incarcerated than white males.[12]

The US is often threatened by natural disasters, and sometimes these are not handled well. When Hurricane Katrina hit in 2005, it was a mess: stores were sacked; there was a shortage of food; and aid, albeit promptly offered by the international community, took a long time to get to the people in need. Some loaded planes had to wait for authorization to enter the US, while others already at American airports were put on hold. There were 1,833 fatalities.[13]

Mexico now has a good record of dealing with natural disasters. The DN-3 plan implemented by the Mexican Army is invoked whenever a natural disaster takes place in Mexico, such as hurricane, floods or epidemics (such as swine flu). The army provides food, shelter and medicines, and reconstructs roads when needed. During Katrina, Mexico offered to help with their trained army, emergency resources and food. They sent 184 men for twenty days, plus helicopters, buses, military vehicles, food and medical supplies. It was the first time since World War II that a uniformed foreign army had set foot on American soil.

In order for systems of security to work, property must be respected. The law must be designed in a way that gives judicial certainty to everybody. It's a simple formula: when the law is well designed, when it is clear and enforceable, the judicial system becomes trustworthy and therefore secure.

Security is reptilian, and it's not just about being secure from invaders or safe inside, it's also about economic, judicial and social security. It's about freedom with responsibility; it's about having a backbone. The qualitative guidelines expressed in the table at the beginning of this chapter can also be supported by hard data. Consider, for example, respect for institutions, a relatively qualitative concept that can be measured quantitatively through corruption statistics and interviews. Also, economic inequality measured by the Gini coefficient can shed light on the degree of cohesion and social welfare within a society, which ultimately tells us more about the degree of security it offers at the level of the individual.

Indicators of a nation's sustainability are also important, such as the ratio of external debt to GDP, national savings, population growth rate and foreign exchange reserves. Having good indicators in these areas reflect a country's ability to face adverse economic situations, such as bond crises or stock-market collapses.

In the environmental sphere, we must also consider threats to individual security that are beyond human control, such as natural disasters. Given that we cannot diminish the risk of experiencing natural disasters, such as hurricanes and earthquakes, we need to focus on the population's environmental vulnerability.

Health and education are also considered in the quantitative analysis for security. When provided efficiently, they can engender survival security (health) and aspiration for a better quality of life (education).

Peacefulness is also measured as part of one's basic personal security. Low levels of corruption, combined with political stability, are important prerequisites for judicial effectiveness, which can help facilitate both entrepreneurship, through the enforcement of contracts, and low crime rates through effective law enforcement.

All these elements are considered in the Human Security Index, but since we are trying to broaden the concept of human security, we have included the *Index of Economic Freedom* prepared by The Heritage Foundation. Protection for property rights, governance free of corruption, and an atmosphere that encourages learning and creativity are as important to modern human security as is a low crime rate. The Security Index is the result of the arithmetical average of the *Index of Economic Freedom* and the Human Security Index. The table on p. 158 provides us with a clear picture of the security offered by each country.

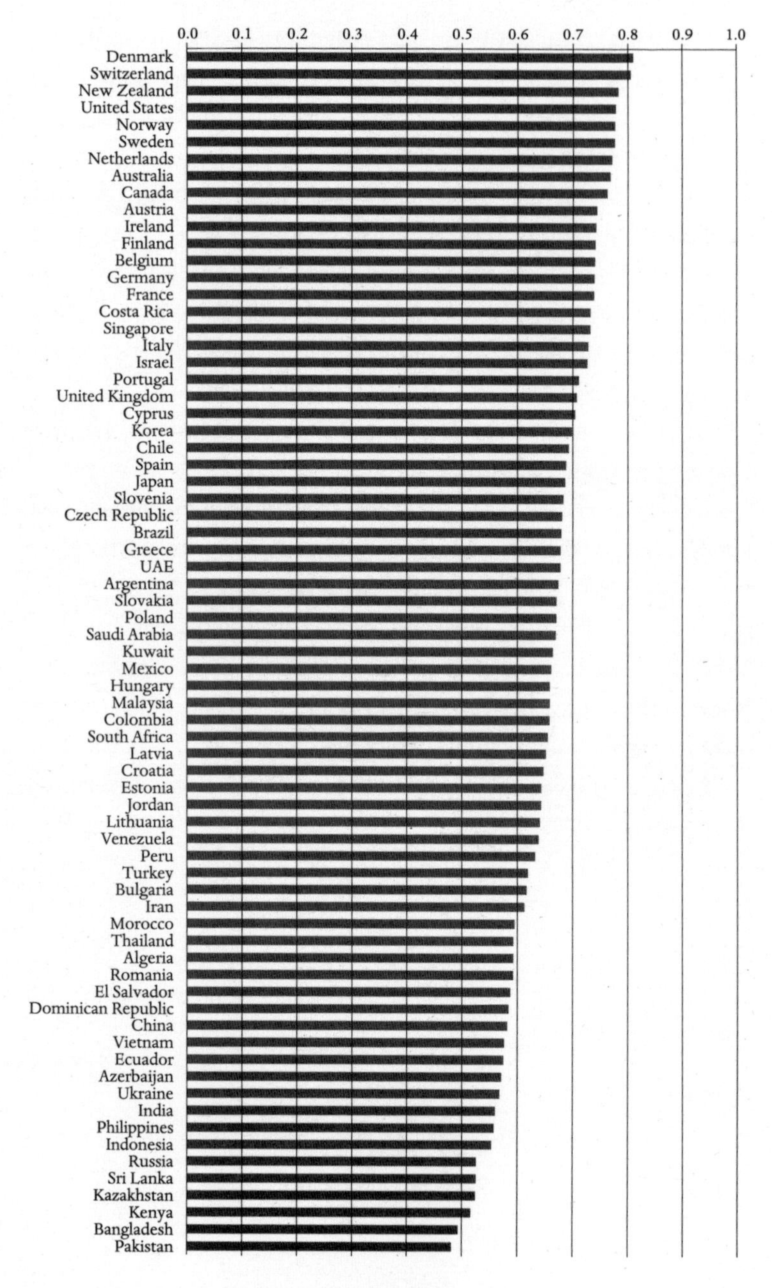
0.0
0.1
0.2
0.3
0.4
0.5
0.6
0.7
0.8
0.9
1.0
Denmark
Switzerland
New Zealand
United States
Norway
Sweden
Netherlands
Australia
Canada
Austria
Ireland
Finland
Belgium
Germany
France
Costa Rica
Singapore
Italy
Israel
Portugal
United Kingdom
Cyprus
Korea
Chile
Spain
Japan
Slovenia
Czech Republic
Brazil
Greece
UAE
Argentina
Slovakia
Poland
Saudi Arabia
Kuwait
Mexico
Hungary
Malaysia
Colombia
South Africa
Latvia
Croatia
Estonia
Jordan
Lithuania
Venezuela
Peru
Turkey
Bulgaria
Iran
Morocco
Thailand
Algeria
Romania
El Salvador
Dominican Republic
China
Vietnam
Ecuador
Azerbaijan
Ukraine
India
Philippines
Indonesia
Russia
Sri Lanka
Kazakhstan
Kenya
Bangladesh
Pakistan
Security Index

11 *Success*

SUCCESS MOVING DOWN ⟶ MOVING UP	
Closed to change	Open to change
Success is acquired through nepotism and inheritance	Success is acquired through hard work and talent
There is life after death	Life is here and now
You stay within your comfort zone	You leave your comfort zone
Karma is a force which controls your life	The only force in your life is your brain
Culture of apathy	Culture of curious minds
Preference for comfort	Preference for taking risks
Ambition is a negative quality	Ambition is a positive quality
Dreams stay in your head	Dreams turn into actions
The bare minimum is satisfactory	People go beyond their basic responsibilities

The relationship between citizens and government is too distant	The relationship between citizens and government is close
The education system is rigid and does not promote creativity	The education system is dynamic and encourages creativity
Poor investment in the science and technology industries.	High levels of investment in the science and technology industries.
No entrepreneurial culture	Entrepreneurial culture
Piracy is common	Property rights are respected
Caste hierarchies divide society and social progress is impossible	The society has no caste hierarchies and everyone is free to move
Beliefs make people think they have to suffer in order to succeed	You only have to work hard in order to succeed
Political connections are what you need most in order to succeed	Talent is what you need most in order to succeed
Fundamentalist values	Open-minded rules
Success is corrupt	Success is admirable
Being rich is a sin	Being rich means you did things right
You are rewarded for following the rules and staying inside the box	You are rewarded for being inventive and having original ideas

Risk is scary	Risk is good
It's about what you have	It's about what you are/know
Destiny is responsible	Freedom is responsible
God dictates everything	Freedom of opportunity dictates everything
Poor countries blame their problems on others	Poor countries recognize internal corruption is to blame

Human beings are obsessed with moving up, growing up, standing up, living up, up, up. We worry about our status, we yearn to be accepted into the top universities, we want to climb our career ladder as quickly as possible, and we strive for the best possible social position. We measure success by our capacity to move upward. Why the hell are we so interested in succeeding? Because humans evolved in small social groups in which superiority was always important. Not only for survival, but also in order to impress friends (and enemies), attract mates, intimidate sexual rivals and look after children. Our social-primate brains evolved to pursue a central goal: to look powerful in front of others.

We want to ascend because subconsciously we all feel that being on top gives us more chance of survival and reproduction. As Robert H. Frank said, 'We come to the world equipped with a nervous system that is worried about status. Our reptilian instincts are constantly asking: Where am I? How appreciated am I? What's my worth compared with everyone else?'[1]

The night Megan met Bradley, she was stunned by his perfection: a well-built, handsome, tanned rugby player who had just won his game against the arrogant French. Not only was he good-looking, he was also successful. If Bradley had been just another pretty face, maybe Megan would have thought twice before cheating on Richard. But, no, Bradley was the whole package. She loved the fact that she was about to hook up with Mr Perfect, the attractive Australian champion rugby player, surpassed only by David Beckham.

The reptilian instinct has an implicit code for success: it looks to elevate itself above the rest. But it's hard to separate success from competitiveness, ambition and even envy. We compete against others, but our harshest critic is often our own mind. We strive to be better than before: it's a battle we face every day that convinces us that life has a purpose and we are making a mark. So what does this tell us about success? The harder you worked for it, the more impressive you will be. That's what we all want to achieve.

$$\textbf{SUCCESS} = \frac{\text{what you } \textbf{ACHIEVE}}{\text{what you } \textbf{EXPECT}}$$

Success varies from person to person, but what matters is how you become successful. In this sense, achievement is crucial, and so are expectations. You can improve your success in two ways: either by increasing your achievements or by lowering your expectations. Countries that focus on achievements are successful even when they increase their expectations. They are used to a work ethic, and to achieving great things. On the other hand, countries that have little ambition can also be successful, with little effort. Which one do you think accelerates mobility more: the one that works hard, or the one that has no ambitions?

The way that success is defined varies between cultures, just as it does between individuals. As far as we know, success cannot be defined only by how much money you have; it's about the amount of knowledge you are able to gather and your capacity to use it. When it becomes more important to have a master's and a doctoral degree, the gap becomes more exclusive and specialization turns out to be a reinforcing measure for success. Money alone is not enough; saying something worthy of attention is a key differentiator. In any case, the criteria used within different cultures determine how success is perceived, and these critera vary from one culture to another. Moreover, they define and shape how a given society will evolve in its social and political forms.

In American culture, where there's no aristocracy, the only proof of success is money and how much you have left over to give to charity. French culture couldn't be more opposite in this sense. In France, having money is bad; it means that you must have exploited people along the way. Success is based on how much of an intellectual or an artist you are, what new ideas you contribute, and how you can transform the world with your ideas. Success is when people acknowledge that you have a fantastic mind. The Pasteur Institute is a symbol of French culture, where being a researcher is prestigious, and where your success is due to your brilliance. You have made breakthrough discoveries that will make the whole world better.

In China, being successful means being virtuous, a belief strongly linked to Confucianism. If you are a leader, you have an obligation to do what is best for your people. You should dedicate your life to the good of the people that you lead. That is success; it has nothing to do with power, property and so forth: it's about the common good. However, because the Chinese highly value discipline and loyalty, they also value hierarchies – more than Americans, for instance. Since status is now more easily acquired than in the past, meaning that it can be *bought*, Chinese actually consume more high-status brands than Americans do.[2]

In the 1986 World Cup, Argentine football player Diego Maradona helped Argentina win the Cup with an un-penalized handball. Argentineans actually celebrate this and call it 'the hand of God'. In that instance, that was success. A culture that celebrates cheating and corruption (as in that football match) is not going up, is it? The mantra in Argentina is, 'If you are opportunistic and savvy, you'll succeed.'

Shades of Colour

If we are talking about success on a family level, the first reptilian dimension is simple: it's about feeding your family, providing them with all the basic needs and resources so that they can survive. The second dimension for success is making sure that they feel good, making sure that they are learning and progressing as individuals.

In South Korea and China we have a very good example of the tension between being caring and being demanding. When mothers are caring but not demanding, people don't move up. In the US for instance, where mothers are more caring than they are demanding, everybody gets a diploma even if they don't deserve it. In Shanghai and Seoul on the other hand, mothers are both very demanding and very caring, thus youngsters are pushed to succeed. It's no surprise that students in those countries score much higher on test results than their American counterparts.

The third dimension of success is moving up. A successful

culture is one where its people can move up. Mexico has been unable to achieve this third dimension because for so many centuries nobody was able to move up, from the repressive Aztec Empire, to Spanish colonization, to an oppressive political system. Benito Juarez was an exception, being the first indigenous president who made his way up from nothing: he is a national hero. The variation in the levels of success between countries is evident in the stark contrast between cities on the border between Mexico and the US: Tijuana v. San Diego, or Ciudad Juárez v. El Paso.

It's easy to observe a person behaving differently on one side of the border than on the other. When Mexicans are in Mexico they litter the streets and exceed the speed limit. When the same individuals cross the border, it's almost magic: they don't throw litter and they respect the speed limit. Americans, on the other hand, cross the border to Tijuana and become party animals – drive drunk, consume drugs and do everything they can't do back home.

The code for success in the United States has always been clear ever since the country was created. Success was on the map from its very beginning. It's in Americans' minds, so it's no coincidence that success is called the 'American dream'. Every American wants this: it's in the air they breathe. The United States never had the barriers to progress that Mexico had. The whole point was to get to America and then strive for the American Dream, to move up. As a result, migration between Mexico and the US is extraordinarily high. Today, there are more than 30 million Mexican descendants in the US, and an estimated 11 million Mexican-born migrants.

People vote with their feet: they move to where they can be more successful. More laws, regulations and taxes don't work as expected at first sight; people just get up and leave if they don't like them. It's the whole reason why the American settlers left England to start afresh. They wanted to move up, so they left; they voted with their feet. If taxes are lower in Florida and Texas than they are in Connecticut or California, people will simply leave the high-tax states. Switzerland has a 22 per cent business tax, while Singapore has a 15 per cent tax and the US a 48 per cent tax. It doesn't matter how patriotic you feel: if your country isn't working for you then you leave, like Gérard

"Hello! What's this?"

Depardieu. It's a competitive world and countries also have to compete for the best talent, and create an environment that makes people feel this is the best place for them to move up.

Talent is allocated according to where it is most convenient. The United States attracted terrific scientists during and after World War II, and it will do so again. The US is the world's biggest economy and *the* birthplace of innovation, although China, India, South Korea, Taiwan and Brazil are quickly catching up. According to a study published in *Nature Science Report*, Boston, Berkeley and Los Angeles are the cities that contribute most towards research in science and engineering. None of the world's top 100 cities that do research in such fields are in Latin America.

Success is related to people being able to move up and feel the rewards of moving up. It is not about being rich; it's about how you become rich. If you look at successful people, we tend to admire most those who are self-made – Bill Gates, Steve Jobs and Richard Branson. Also, the more tragic their background is, the more impressive their achievements – look at Oprah Winfrey and Stephen Hawking. When successful people have inherited their wealth or had an unfair advantage, they are no longer so impressive. Take Paris Hilton as an example. Do you admire her? Probably not. What she does is simply spend her family's fortune. Another example is Mao's granddaughter who in 2013 appeared in *Forbes* magazine as one of China's top 100 millionaires. Although rumour has it that she actually became rich after marrying her current husband Chen Dongsheng, it is still ironic that she is a millionaire while at the same time she supports the ideals of her grandfather.

Success has nothing to do with being rich: success is more like a journey, and being rich is a result. Success is what you *do* to move up.

We want everyone to have the opportunity to move up.[3] That is, everyone should have the same opportunities of climbing the ladder of success. Yes, this means that there will be winners and losers, but that should be nothing to do with race, gender or ethnicity. Rather, it will be down to pure merit: you work for it.

The purpose of our index is to predict which cultures create a space where their people can move up and which don't. Can you move up in Russia? Not really, because most of their talent is leaving, and the population is shrinking. On the other hand, there are aspects of Brazilian culture that help people move up: for one, racism is less prevalent than in other cultures – it's OK to be successful regardless of your race, and it is also OK to move up and be successful. The only issue is that once you are rich your safety is at risk. In a feminine culture like Brazil, success is less aggressive.[4]

Offence is the Best Defence

For centuries, the notion of success was defined by the possession and occupation of territory, and invading your neighbours. William the Conqueror won the Battle of Hastings because his soldiers had better stirrups than the English, and was able to invade England and become King. European countries invaded and conquered each other for centuries through the use of violence, and that was the measure of success back then.

Today, the battle is different. Competition is between individuals and between companies representing their employees. Coca-Cola versus Pepsi is a battle of who has the best ideas, advertising, strategy and promotions, but also about who has greater territory. It is about who has more resources and clients to ensure their survival in the future.

A way to judge success at the individual and group level is to look at whether people develop in an open or closed system. A closed system is never going to allow individuals to move up. It assumes that the rules and the environment are going to stay the same, even though both are always changing. A closed system will place obstacles all the way: barriers within the market, illegal practices inside the government, and expenditure on non-beneficial areas of society. On the other hand, an open model will allow individuals and groups to move freely without finding impediments to their growth and development. It is more possible to comply with the objectives set by the reptilian, the limbic and the cortex. It will be easier for people to find healthy food and to survive to enjoy their emotions and, in consequence, to see further into the future, when they will retire and enjoy the results of their hard work. An open system will be best for individuals to learn and enjoy to their fullest capability.[5]

We can talk about success and winning using two models. There's the sports model, where the rules are clear and there is a single winner. In soccer you have a 90-minute game and a referee; the rules are clear and stay the same throughout the whole match. But the world is not like that; it's more like the second model, the business model, in which there is never a declared winner in the end and the rules keep changing constantly. In business there are no defined sets of rules, and what rules exist are different in every culture. American businesses make that mistake when they venture abroad, thinking that other countries are going to play by their rules.

In Japan, it is common to hand your doctor an envelope with cash as a gift when you go and see him. Some cultures may perceive this as a bribe, but in Japan it gets you what you need. In France, in order to hire the top talent, you need to have lunch with the candidates first. You need to sit down with them in a casual environment in order to read and decode them. Though you may need an expert accountant, you want to see if they know how to use a knife and fork properly. It all depends on the rules of the culture.

In Japan, success is perceived to be down to the group, not the individual. If you promote one person, you have to promote everyone else in the team. There is a Japanese proverb, '*Deru kui wa utareru,*' which means, 'The stake that sticks up gets hammered down.' The Japanese are very group-oriented, and this is good when creating an environment of zero corruption, because as soon as an issue pops up it is punished, either socially (no honour for the corrupt man) or judicially (prison). However, when it's meritocracy's turn, people are also punished for standing out. The Japanese are so group-oriented that there is little room to step out of the box: the culture makes it hard to be different from the norm. Innovations are usually incremental, rather than major breakthroughs. In such a culture, individual success is regarded more as a burden than as a goal; but, when it is about succeeding as a group, they are the best.

In Mexico and other Latin American countries, there is the concept of '*dedazo*', which describes a common phenomenon whereby the sitting president appoints his successor and other officials by 'pointing his royal finger' at them (the term is derived from the Spanish word for finger, '*dedo*'). The succession process known as *dedazo* or 'The Institution of the Finger' is a measure of favouritism and discretional decision-making to favour someone without taking merit or talent into account. There's no meritocracy under *dedazo*.[6]

The way that the indicators of success vary from one culture to another is fascinating. For example, the British, being born British, already feel superior so they don't need to prove anything. You can wear your grandfather's old polo boots, which shows that even your grandfather played polo. But if you bring new boots to a game, you're revealing that you're nouveau riche. If you tell a British person that they have a beautiful home they may get offended, assuming that it must be obvious that they would have a beautiful home. 'What, you expect me to have a horrible house?'[7]

One measure of success is acceptance. You have to choose who you want to accept you. You might take a risk even though you might be rejected. If you want to move up, you need to understand different social levels and their nuances. You need to read all the rules, to understand them in order to signal your success and even to decode what level you want to move up to. To live in some fancy American condominiums, you need to submit an application that all the other tenants review before accepting or rejecting you. It's about *who* you know rather than *what* you know.

We hunger for success, and success is sometimes demonstrated by what you own: your car, your suit, your handbag, your glasses, your phone, even your dog. Shirts with the logo of a man on a horse supposedly advertise someone's social status. We think we show status with the things we have and use, regardless of whether the status we show is true or not in terms of *actual* power. We think that by having a Hermès Birkin bag, something will change. Why do we spend between $15,000 and $20,000 on a bag that serves the same function as every other bag, and why is spending so much sometimes considered tacky? And why do some groups appear to show off more than others? Thorstein Veblen gave us a theory of conspicuous consumption in the nineteenth century that has begun to be proved in experiments.[8] One such study showed that people in a powerless position were willing to pay more for goods associated with high status than those associated with low status. In another, people were more generous when someone was wearing clothes from a highly respected brand. The lack of power compensation mechanisms suggests why blacks and Hispanics, groups whose power is limited relative to their white peers, spend more on goods that are considered conspicuous than whites of the same income level. There are subtle mechanisms to show one's status – sometimes showing one's status can mean not being obvious, like when the more expensive an item is, the less branding it has. And even for sex: how many men think of having a Ferrari to attract more girls?[9]

Another measure of success is related to envy. Your success is always relative to someone else's; you can't envy yourself. We constantly compare ourselves with our peers. At a school reunion everyone is eager to know how they are doing in comparison to their old classmates. Our reptilian needs to compare itself to others, and constantly challenges us, making us feel envious. This is a mechanism for doing something about the threat of your neighbour's superiority.

$$\textbf{ENVY} = \frac{\text{what your neighbour has}}{\text{what you have}}$$

Just as with success, envy can be measured using the formula above. When what your peers have is more than what you have, you feel envious. Conversely, when what you have surpasses what your neighbour has, guess who envies whom?

But don't get confused. The point is not to learn how not to feel envy, since we can't help it – it's a natural survival mechanism. However, cultures can make such feelings work for or against their society, and there are two ways in which cultures can channel envy: trying to get what your neighbour has and more, or trying to make your neighbour fall and make things 'even'. In other words, you can wish for what he has, or you can wish he had what you have. When you envy someone, you should take more exercise, work harder or study more to become faster, stronger, richer or more cultured.

The other side of envy is exemplified through the popular saying, 'Dear God, if you can't make me skinny, please make my friends fatter.' Minimizing the other's achievements, or even making it impossible for them to become better, is harmful for mobility. A world governed by this philosophy won't evolve. The 'crab mentality' is widespread in Latin America and the Philippines. The name comes from the idea of crabs trapped in a bucket. Those who try to get out are pulled back down by their peers, and thus no crab can ever leave the bucket. There is no mobility; no crab ever escapes. The bottom line is that envy as ambition is a moving up factor. Envy as *schadenfreude* is a complete mobility killer.[10]

Daron Acemoglu and James Robinson are convinced that what makes the difference in the wealth and well-being of a country's population is the nature of its institutions.[11] Institutions are cultural expressions of societies. Extractive institutions are designed to extract the most resources possible from a country. These types of institutions benefit an elite, which is often the same as, or is closely related to, the political class. Contracts are awarded to friends of politicians, or to companies owned by politicians.

In cultures with extractive political institutions, the goal is not to found a company and compete in the market, but rather to be friends with the people in power so they give your company contracts. This is the model of Latin America and many Asian nations, in particular, including China. The media in these countries self-regulates in order not to say what the government doesn't want them to say, since saying otherwise can jeopardize government officials' sympathy. In such a society, success is acquired through nepotism, friendship and inheritance instead of through hard work and talent. It's not only about politics; this approach permeates every sphere of economic transactions.

China has managed to grow rapidly with extractive political institutions, and, as Acemoglu and Robinson state, although growth is possible it's not sustainable, nor desirable. China opened the doors to foreign investments and technology, so Chinese economic institutions are more inclusive, but political institutions are still repressive and restricted to an elite in the Communist Party.

When political institutions become more inclusive, economic institutions flourish as a result of competition. China must undergo deeper transformations to protect property rights, promote a process of creative destruction, and open its political system. This is the only path to sustained economic growth. Adoption of foreign technologies can't last forever, and, once China has caught up, its growth will slow down dramatically unless the Chinese develop more inclusive political institutions, make deep changes in their attitude to property rights, and foster a creative destruction process in order to create their own technology.

Not All That Glitters Is Gold

Success is not just about money. In Saudi Arabia you have rich, spoiled second-generation kids who have everything and therefore don't care about making anything of themselves. They don't care about success.

There is no success without an evaluation of risk and reward. Risk cannot be understood without net reward

(cost–benefit analysis). I'm going to risk a great deal if the reward is big enough. The pioneers who came to America took a big risk, but their potential reward was to create a new world, to be free, to have a new life, so they found the risk worth taking. If the reward is low then people would much rather just live comfortably. Risk is short-term and therefore reptilian, whereas reward is long-term and more cortex.

There are always real risks, even physical risks, but the greatest risks are those we create in our minds. The real limit for moving up is not an evaluation of external risk, but a program, a pattern, invented in your mind that tells you that you cannot do it. Religion plays a key role here. The Bible says that it's harder for a rich person to pass through the gates of heaven than for a camel to pass through an eye of a needle. On the other hand, the Protestant faith encouraged the development of capitalism, encouraging people to save and accumulate wealth. What matters is not just the text, but how you read it.

On an individual level, if I don't take a risk am I going to blame myself? Does it mean I'm just not good enough? Am I going to find an excuse to make me feel better about myself for not taking the risk? Blame others and not myself? One of the key elements of moving up, a key element in all self-help books, is taking responsibility for your own life. No excuses: you are responsible for your life, for your destiny. Cultures that place an emphasis on God, karma, religion or horoscopes are cultures that don't help you accept personal responsibility.

While it's true that we are never completely in control of any situation, this is not due to karma or a certain alignment of the stars. Rather, it's because of the genes we have, the good or bad luck we are surrounded by, the culture we are raised in, and the experiences or traumas that come our way. We do not choose any of these: we are born with certain genes, in a certain culture, and events sometimes just happen around us, whether we want them to or not. However, that doesn't mean that we can wash our hands of responsibility and attribute everything to chance, karma or the horoscope. We still have some degree of freedom and we are responsible for our actions.

There are cultures where the notions of risk and reward are very clear. Those are the cultures where people are moving. People will try to move away from a repressive culture, and towards a culture that helps them move up.

We are talking about sensible risks, not reckless risks that could lead to another subprime-mortgage crisis. We are talking about entrepreneurship.

Start-ups are the future. They are the engines of sustainable growth.[12] Entrepreneurs are willing to take the risks of innovation in exchange for a reward. This is how it works. So countries that impose a high tax rate on start-ups are not really helping their society. Who is going to take the risks if there is no reward? It would be stupid to even contemplate it.

You might think that lowering taxes would make entrepreneurs richer, and widen the inequality gap. Possibly, but, as Robert Cooter and Aaron Edlin have demonstrated, the gains in welfare outweigh the losses in equality for everybody, even the poor.[13] In short, it's best for everybody to encourage innovation rather than inhibit it. China encouraged entrepreneurship after the disastrous Cultural Revolution, and the result was that inequality grew, but the Chinese people, including the poor, are so much better off today.

Conservative societies that don't encourage creativity and technological change will fall behind. Trading apples and pears no longer creates wealth. Today, wealth is created by trading ideas. Additionally, nobody loses through the exchange of ideas. It's simple mathematics: one idea plus one idea gives us two ideas, and you can't eat ideas, but you can eat apples, and apples will eventually disappear.[14] Sure, apples are food, and food is needed for survival, but who doesn't need computers today?

During his visit to China around 1980, president of the World Bank Robert McNamara discussed with Deng Xiaoping the opportunities that China had to get involved with the west. The World Bank would send a team to make a study to define a strategy for China to follow. Deng told McNamara that ideas were much more important to him than money.[15]

One way of promoting creativity is by exposure to the unknown. Travelling to different cultures and trying to understand them results in eye-opening experiences that improve our cultural intelligence, and make us better equipped to face new situations and problems. Between 1978 and 2008, one million Chinese students went abroad, of whom a quarter came back to China. Many of the economists who designed the economic policies of today's China were trained in the United States, sponsored by the Chinese government and the Ford Foundation.

Some countries fear that if they adopt such a policy they will suffer a brain drain: all the country's bright minds will flee never to come back. Actually the latest data on patents and innovation suggest the contrary. Diasporas in middle-income countries serve as a bridge to the home country's skilled talent. Researchers abroad often make contributions with their home-country peers. A report prepared by the World International Property Organization (WIPO) found that expatriate inventors from middle-income countries were more likely to share patents with inventors residing in their home country. That is, they were still committed to their homelands. The countries with the highest ratios were Ukraine and Mexico.[16]

In February 2014 the governments of Mexico and the United States launched the Bilateral Forum on Higher Education, Innovation and Research, which intends to tighten bilateral cooperation on human capital with the United States. The goal is to send 100,000 Mexican students to American universities, and 100,000 American students to Latin America, half of them to Mexico. This policy is not new; other countries have made the same commitment with positive results. In the Meiji era during the second half of the nineteenth century and the early twentieth century, Japan was modernized from a feudal country to an industrial state, and education played a central role. Education became public and universal, and, in order to catch up with the west, professors were brought from abroad to teach in Japan and Japanese students were sent to the United States and Europe to study. Japan ended up having state-of-the-art human capital in both schools and government. In Singapore, Lee Kuan Yew, the so-called father of Singapore, also adopted a policy of having the best human capital in the world, which meant sending students overseas and welcoming educated foreigners. Schools are now bilingual (English is mandatory, and the mother tongue is taught for literature), which enables them to compete with the developed English-speaking world. Moreover, Singapore is now on the other side of such exchanges: they are receiving an increasing number of international students. The government has even launched an

"We've considered every potential risk except the risks of avoiding all risks."

initiative to make Singapore a 'global school' and increase the number of international students in Singapore to 150,000 for 2015. It is these kinds of visionary policy that ultimately enable its population to move up.

Moving up means satisfying the three brains in this order: the reptilian (through the provision of food, shelter and safety), the limbic (through love and happiness) and the cortex (by buying a home for your retirement).

Money is the worst way to reward success because it doesn't last and you'll forget about it quickly. The best reward is a new identity – a new title perhaps. Napoleon was really good at giving titles to those who did a good job, to create new potential identities for them.[17]

Does your country have a ritual or an institution that hands out rewards? If people move up, how do they know that they are moving up? In some cultures it is keeping up with the Joneses and conspicuous consumption. In Britain it is getting a title, like Baroness Thatcher, Sir Paul McCartney and Lord Sugar. If a culture doesn't reward you, then why bother?

Success varies between cultures. It is very difficult to have one set of criteria to measure success worldwide. Different notions of success can be linked to different notions of what is perceived as fair and just in terms of competitiveness. Throughout human social development, there has been one central debate: who has the right to decide? Who can make the best decisions? Those who are successful? Or is everyone competent to make a decision? This is where the concept of meritocracy comes in.

The Economist published an article about the formation of a seemingly 'new' meritocracy in the United States. From who is capable of decision-making to who gets access to a better quality of education, the meritocracy debate is increasingly important today due to the widening gap between the poor and 'the clever rich [who] are turning themselves into an entrenched elite'.[18] Social mobility is currently slowing for those who do not have access to high-quality education. While the rich elite invest in their children, as they have the right to do, the educational system ditches because it adds up to an already privileged elite.

Where meritocracy fails, our greatest risk is a type of plutocracy, where the wealthy elite rule and remain removed from the troubles and issues facing the masses. However, in a place where meritocracy rules, we see the opposite: the correct allocation of talent and resources naturally falls into place and individuals can truly look up and strive for success.

Data

The data considered for the index reflects the cultural traits from the beginning of the chapter. For Success we used the Global Competitiveness Index prepared by the World Economic Forum for Success, which accounts for many factors that help a society to succeed. Property rights, lack of corruption, business ethics and governmental efficiency are variables that promote entrepreneurship, and therefore individual success.

General infrastructure for transport and electricity enables people to do business more easily and more inclusively. Of course, a positive macroeconomic outlook is particularly important for success (it will affect investment decisions, for example). Education and health and their quality are necessary conditions for a person to move up, especially in undeveloped countries. Besides, education is the prime engine of innovation.

Tax rates and market efficiency are also considered in the Success index not only because they are measures of competitiveness, but also because the way taxes are designed can either promote or inhibit movement within each society. In order for markets to work properly, not only do they need an appropriate tax structure, but there also needs to be financial inclusiveness. This can come in many forms: venture capital availability, flexibility in the labour market (both for individuals and corporations) and access to essential technology such as the Internet and mobile phones.

Now that we've described the conformation of the C^2 value in the previous section, and defined the bio-logical rankings of the Four S's in this section, we are ready to look at the R^2 Mobility Index and our results for the seventy-one countries we have analysed. Is your country moving up?

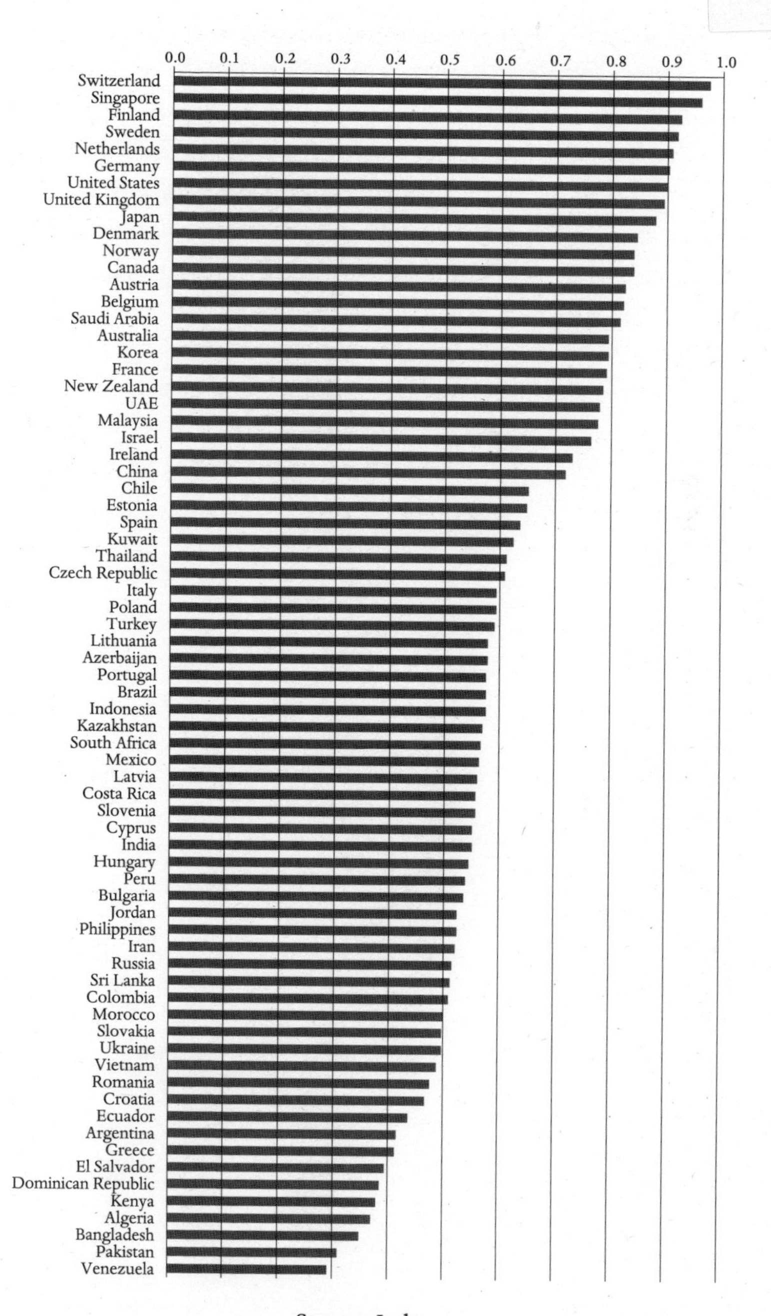
0.0
0.1
0.2
0.3
0.4
0.5
0.6
0.7
0.8
0.9
1.0
Switzerland
Singapore
Finland
Sweden
Netherlands
Germany
United States
United Kingdom
Japan
Denmark
Norway
Canada
Austria
Belgium
Saudi Arabia
Australia
Korea
France
New Zealand
UAE
Malaysia
Israel
Ireland
China
Chile
Estonia
Spain
Kuwait
Thailand
Czech Republic
Italy
Poland
Turkey
Lithuania
Azerbaijan
Portugal
Brazil
Indonesia
Kazakhstan
South Africa
Mexico
Latvia
Costa Rica
Slovenia
Cyprus
India
Hungary
Peru
Bulgaria
Jordan
Philippines
Iran
Russia
Sri Lanka
Colombia
Morocco
Slovakia
Ukraine
Vietnam
Romania
Croatia
Ecuador
Argentina
Greece
El Salvador
Dominican Republic
Kenya
Algeria
Bangladesh
Pakistan
Venezuela
Success Index

THE R^2 MOBILITY INDEX

12 *Is Your Country Moving Up?*

Ever since man first showed off his genitals to women in the middle of the jungle to signal his availability for coitus, we have always been trying to show off, to be better, to move up. Countries compete for the best GDP and weapons of mass destruction; cities compete for the best transport system or tallest building; teenage girls compete over who has bigger breasts and boys compare penis size. We are competitive by nature, and our objective is always to be better, faster and stronger.

As a popular saying states, 'Statistics are like bikinis: what they show is pretty suggestive, but what they hide is essential.' The latter is why most prosperity indexes fall short. Of course, it's vital to know which country produces the most goods and services, which is the most environmentally sustainable, which is the most egalitarian, safe, competitive and even 'happy'. But there is no relevant indicator currently that measures a country's prosperity based not only on the outcome, but on the conditions: those factors that influence a country's mobility.

The Gross Domestic Product index measures the total production of a country in terms of money, without taxes and quotas. It is used to compare economic advances among countries. The problem is that it does not measure citizens' quality of life, nor their cultural or educational advances. It only measures the amount of production (from agriculture, services, factories, small business and so on). So it's impossible to determine whether people living in a country are healthy, educated or happy.

The Human Development Index, an indicator created by the United Nations Human Development Programme, measures advances in human development, including economic, health and educational aspects, but it does not show how people use their values and cultural baggage to increase their own development.

With these two indicators it is possible to identify countries with high levels of production, health and educational standards, but we cannot be certain whether people living in those countries are able to move up – if they are able to study or work where they want to, or achieve their dream job.

Other indicators have been developed in order to measure fuzzy concepts such as happiness or well-being. In this category we find the Well-Being Index in the US, used to assess the health and overall well-being of its citizens. This includes physical and emotional health, healthy behaviour, working environment, community development and so on. But so far it has only been used to measure the well-being of US citizens.

The Happy Planet Index combines environmental impact with well-being to measure the environmental efficiency with which, 'country by country, people live long and happy lives'. It rates countries in terms of life expectancy, life satisfaction and ecological footprint, but it misses the economic, political and cultural components.

In summary, there are many indicators that evaluate several aspects of life, society and the economy within a given country. But even with all the available information, we could not find an indicator that showed what makes people move up. In other words, an indicator that reflects the cultural and the bio-logical elements that influence mobility, those factors that we have revealed in this book. Thus we had to build our own indicator: The Rapaille–Roemer, or R^2, Mobility Index is an indicator that explains how societies around the world are being bio-logical and on-code with their culture in order to move up.

Basically, according to the UN, indicators are tools for defining objectives and impacts; they are also measures to verify changes or goals. You can use an indicator to verify your position in comparison with another person or country. That is why we were intrigued to find out what makes us move, what makes countries move, and thus develop an indicator that helps us compare the conditions for moving up among countries.

How Do You Read the R^2 Mobility Index?

The R^2 Mobility Index proposes a new way of understanding mobility. What we are really interested in is measuring the factors and conditions that allow individuals within a culture to move up. The index does not measure mobility. Testing for mobility is particularly difficult since there simply isn't enough data to do it, at least not for all the countries considered in our index. Instead, the index shows how each country is fulfilling the conditions for moving up.

Measured on a scale of 0 to 1, a low value indicates that the society has few or weak policies that promote upward social mobility, whereas a high value indicates that a country instils strong policies that promote upward social mobility. To state this more clearly, if a country provides optimal conditions for moving up, its index value is equal to 1. If, on the other hand, its index value is zero, then there's no upward movement whatsoever: it just remains static. None of the countries in our sample falls at either extreme; all of them are between 0.22 (Bangladesh) and 0.85 (Switzerland).

The R^2 is an aggregated index, comprised of the bio-logical variable (made up of the Four S's) and the cultural variable, the C^2. Each S goes from zero to one, following the same logic as the R^2; that is, zero is the worst, and 1 is the best value possible. The same rationale applies to the C^2, which also ranges from 0 to 1, zero being a culture that prevents its constituents from moving up at all, and 1 being a culture that encourages its individuals to reach the top.

$$\mathbf{R}^2 = \frac{C^2 + \text{Bio-Logical}}{2}$$

The R^2 indicator has two main components: one qualitative and another quantitative. Since it is hard to assess culture, we have defined a qualitative methodology (the 'five critical moves' and the 'third unconscious') and developed it to produce one number, the C^2 variable. This variable considers many cultural characteristics within four basic axes: freedom, innovative spirit, openness (including sexual attitudes) and meritocracy.

For the C^2, we tried to uncover the hidden meaning of crucial aspects that affect mobility such as freedom of choice, innovative spirit, gender equality, security and a drive to succeed. Cultures are complex, and understanding why certain cultural traits arise, proliferate and survive is not always clear-cut.

We evaluated the seventy-one countries and their culture codes according to our 'five critical moves' and 'third unconscious' theories, giving them a C^2 value for the R^2 Mobility Index. One thing that is worth emphasizing is that, in order to identify culture codes, we cannot judge cultures from our point of view. China has never had a democracy, so we cannot assume that it's necessarily a bad thing that they don't have a democratic government. However, what we can do is judge a culture through their own culture codes, and then see if they are on-code.

Once the qualitative part, the C^2, is understood, we can move to the bio-logical value, which is the quantitative part of the index. The bio-logical factor is composed of hard data categorized into the Four S's: survival, sex, security and success. These are the bio-logical factors that influence upward social mobility, taking into consideration our human nature, instincts, innate needs and the capacities essential for moving up. These S's are used as a benchmark for understanding how your culture helps you move up. Does your culture help you survive? Do both sexes have equal opportunities? Does your culture make you feel secure? Does your culture help you succeed and let you feel the rewards of your success?

The bio-logical, or quantitative, part is composed of hard data: it does not contain subjective opinions. It includes measures of wealth (GDP), inequality (Gini coefficient), women in political positions, competitiveness, climate and so on. We measured Survival based on the Where to be Born Index and a country's expenditure on health and education. For Sex, we used the United Nations' Gender Inequality Index. For Security, we used the Human Security Index; and, for Success, the Global Competitiveness Index. However, this hard data is a reflection of other cultural features. What we are more interested in explaining here is the cultural characteristics that give birth to this hard data.

In order to move up, our survival needs beyond basic shelter, food and water need to be met. When a country satisfies the sex variable, it means it is well aware of our instinctive need to reproduce and the innate differences between men and women. Therefore it is important for countries to have policies that support the reproductive needs of the population, as well as to ensure the rights and personal fulfilment of both genders.

Security from threats and safety within the safe zone is also an essential variable to consider with regard to upward movement for a country's population. Lastly, but probably most importantly, the drive to succeed is inherent in all human beings, therefore it is important for a country to create the necessary conditions for their population to succeed.

Which cultures are best at moving up? Firstly, the culture must satisfy the basic reptilian needs of its citizens. Having a culture that encourages the reptilian with no restrictions is not what we are looking for. Imagine a culture that encourages killing your neighbour when there is a shortage of food. That is a very reptilian culture, but do you think such a trait could drive a society to move up? Not really. On the other hand, a culture that denies competition for scarce resources is also out of touch with reality. Rather, a culture that understands that we have such reptilian impulses and therefore creates a cultural framework to allow us to express it in the best way possible is a culture that is moving up.

Secondly, we have to understand what success means within each culture. One of the key elements of American culture is the 'comeback kid' archetype. America believes all of its citizens have a bright future, and they can successfully integrate their immigrants because, in essence, America is a country of immigrants. It's a culture that you can choose to become a part of. If you learn the language and integrate yourself into American culture, you are American. People die crossing the border to get to the United States, they take risks to become American: that says a lot. We would go as far as to say that new immigrants are even more American than those born and raised in the United States because they *choose* to become American.

The United States has a culture that encourages moving up, and a 'keep moving' spirit. There is a reason why Facebook and Apple were created in the United States. The culture rewards risk-taking. American culture is an adolescent culture – Americans are rebels. They don't hesitate in *doing*.

France, on the other hand, is a more pessimistic culture. The French think all the time, they criticize everything and everyone, but they do hardly anything about it. Even their education system was invented by Napoleon and hasn't undergone any major changes since. It's a very cortex culture, but because of this they just end up frustrated because everything is wrong, yet they refuse to do anything to change it. In contrast with the United States, where the culture is oriented towards action, to be practical in France is considered vulgar. But attempting new things is key for moving up.

Whereas the French are busy thinking and making laws to practically ban hard work, Americans are busy creating, producing, making mistakes, trying again and making wealth. The US ranks third in the R^2 index, and France eighteenth. It's a huge difference for a developed, sophisticated country like France, especially considering that its neighbour, Switzerland, ranks first! Both countries speak French, they share geography. Why do they differ so greatly?

With a maximum 24 per cent business tax rate in contrast to a 34.5 per cent rate in France (there was an effort to set the top marginal rate at 75 per cent but this was ruled unconstitutional), Switzerland has a strong, open economy, where property rights are respected and enforced. It has an average tariff of zero on foreign trade, and they have a well-regulated banking system (banks are what Switzerland is famous for). They manage to maintain an environment that encourages innovation – they have the highest patent application rate per inhabitant in the world. Their education is world-class at all levels. They encourage higher education just as much as they encourage apprenticeships, which enables them to have an unemployment rate of less than 3 per cent, while France's hits double digits. The focus of the French economy is on creative destruction through innovation, so it's not surprising that Switzerland came out on top.

Those countries that are moving up are those that are at the forefront. What gets them there? They are open systems that accept diversity, and don't base their principles on dogma. But, not only that, they integrate differences and challenges from others.

In Singapore, there is a mixture of ethnic backgrounds, and Lee Kuan Yew managed to integrate them completely. Indian, Chinese, Malaysian and European descendants, as well as Buddhism, Hinduism, Islam and Christianity, live together peacefully. They even celebrate a Racial Harmony Day on 21 July each year.

In India, on the other hand, there is no integration. People live separate realities because of the caste system. You belong to a caste and you remain there until you die. They believe in reincarnation, so moving up in this life is not possible. But, according to your karma and what you go through, you might move up in the next life. So there is no real revolution in the system; a real revolution would be a revolution between two castes. Separate realities aren't supposed to interact. Moving up is just not possible. Legally they pretend that this isn't true.

You can move up in one caste, you can be a poor dalit and become a rich dalit, but still, real moving up would be moving from a lower caste to a higher caste. This is their cultural shadow, where a very community- and family-oriented culture can at the same time be so segregated.

In Singapore, a meritocratic system is strongly encouraged in school. There is a great emphasis on education and moving up there for several reasons. You are going to be what you put your mind to and, because of that, you are going to move up. What is good about Singapore's system is that you are rewarded for your hard work. There is a discipline in learning that goes back to the Asian perception that 'you don't just learn information, you transform yourself'. Moving up as a person is not just about making more money, it's about transforming your character, becoming stronger from within.

Singapore's education philosophy goes beyond passing tests. There is an ethic behind it: there is no corruption in Singapore. It's not personal morality that stops a Singaporean from cheating; they don't cheat because they know that the only person they are cheating is themself. It's a very simple system with clear rules and clear rewards. Everything is about helping you to succeed. It is a business-friendly country with few regulations (although the regulations that exist are strongly enforced) and low taxes, and there is a predictable and dependable legal system, making it a good environment for innovation.

Finding a suitable taxation scheme could also be a key to movement. Sweden's income tax rate exceeds 50 per cent, one of the highest rates in the world. But at the same time they also have one of the lowest tax rates on businesses, making it cheaper to start a business in Stockholm than in Paris or in New York City.

Low taxes for entrepreneurs translates to fostering

innovation, inventiveness and creativity, an important factor we considered when determining the C². Japan stands out for its innovative environment. The Japanese are hard-working people. Even the way they go on strike would look odd to us: they overwork. They overproduce in order to lower the price of the goods and put pressure on their bosses. Their advantage is that they know how to work as a team.

The cliché is that, as individuals, the Japanese aren't the brightest, but as a group they are the most intelligent. They have a group-oriented culture. During the 2011 tsunami that hit the coast of Japan, there was no looting but rather everyone helped each other because as a group they are very intelligent. For them to move up is about moving up as a group. But that also means that individuals are not supposed to move up in Japan; that is against their culture. Despite their know-how in teamwork, the Japanese are not very good at standing out as individuals. It is actually difficult to move up in Japan. It's good for a culture to be demanding, but not too demanding. This turns out to be stressful.

More or less the same happens in Russia and Mexico. Russian culture is about 'suffering' – that is their code. It's about how much you suffer, almost in a masochistic sense: they enjoy suffering as a strong identity builder. It's a nostalgic and passive sort of suffering. When old Russians get together at a bar, they talk about those days when they suffered so much. So how could there be a culture for moving up, for wanting to be successful, if success means suffering? 'I suffer, therefore I am.' This is not very helpful for mobility, is it?

Brazil has deep problems with corruption, inequality and education. Half the adult population didn't finish secondary school. In Russia, 90 per cent of their adults went to school. Life expectancy in Russia is lower than in Brazil, which makes

Russia a difficult country in terms of Survival. Although Russia has taken steps towards a market economy, it's not fully there yet, and they need to work on their institutions to ensure both judicial certainty and political competition.

For foreigners, the stereotype of young Russian women is beautiful models wearing miniskirts and high heels. This is because it's these kinds of Russian women who managed to find an opportunity for moving up. However, documentary film shot in Russia shows women in towns with a cloth wrapped around their heads. Russian women are trapped in one of two realities: the miniskirt and high heels reality, or the babushka reality. Moving up for Russian women is about moving out. If you don't move away you become a fat babushka.

In Mexico, there is an age-old concept that still prevails, which is 'endurance', or '*aguantar*'. Mexicans are accustomed to endure: to put up with their job, to hold back tears, to take all the adversities of life as they come, and to do nothing about it. This is about enduring terrible suffering with incredible strength. The result is a very strong culture, but only to endure pain and difficult times. Running away from adversity is seen as a sign of cowardice rather than creativity. They find it hard to think out of the box; and those who do are perceived as weak. The result is no innovation, a mobility killer.

However, Mexico is now breaking the mould. The government passed an energy reform that liberates Pemex, the state-owned oil company, from focusing purely on petroleum to focus on what it is best at. Pemex was a synonym of nationalism and homeland; to make any changes with Pemex would mean to sell out the country. Fortunately, Mexico was able to break the spell and, along with other important reforms (fiscal, credit, education, telecommunications, and economic competitiveness), Mexicans are fighting their demons. Today, Mexico has become a freer country for doing business than Brazil, with lower business taxes, which results in lower unemployment.

Argentina has taken a step back. The nation that was once rich is now stuck with a bad government with excessively controlling policies (like seizing control an oil company), and is now going backwards. Inflation is as high as 10 per cent, public debt is rising, unemployment is higher than in the rest of Latin America and higher tariff rates have been imposed (currently 5.6%, whereas in Chile it's 4%, in Mexico 2.2%, in the US 1.6%, and in Singapore 0%). Argentina needs to catch up with the rest of Latin America via good governance rather than populism, import controls and government interference. Argentina is culturally rich, home to many immigrant families since the beginning of the nineteenth century, and they need to take advantage of it.

In a cultural way, the United States is doing well – much better than Russia, France and India – but not so well in terms of security. Whereas the US is one of the top countries in our C^2 ranking, in the bio-logical ranking it takes 14th place due mainly to the security component. The United States faces economic inequality and great divergences among races, where you are more likely to go to jail if you are black. There is a significant gap in education, income and even health between white and black Americans.

In the United States, public schools receive money depending on the support of the district. If your house is in a wealthy school district, schools will be better because the community has more money to support them. If you happen to be a child growing up in a tough neighbourhood, your school district will probably be poor too, which means less money than the richer areas, and that eventually translates into meagre educational achievement. This results in structural inequality since it's the richest kids who have access to better public schools.

Canada does better than the US in Sex and Security. And this is reasonable since Canada has more women holding seats in their parliament than the US. The US has a higher teenage pregnancy rate than Canada, which reflects their poorer sex education. There are also more women who are part of the labour force in Canada than in the United States.

The strong militarization, terrorism threats, income inequality and justice inequality explain a good part of the US's lower scores in Security. Take the incarceration rate for example: whereas in Canada it is 118 per 100,000 people, in the US it is as high as 707 per 100,000!

It is no surprise that the best countries in the security component are Nordic countries. They have high taxes to ensure a high quality of life for *all* their inhabitants. Denmark, Norway, Sweden and Finland are at the top of our table along with Switzerland. It isn't surprising too that the worst countries for security are those hit by poverty: Kenya, Bangladesh and Pakistan. They are also countries with high levels of corruption and weak rule of law. Venezuela, for example, is in 69th place. This is a country that has profoundly crystallized 'solutions'. This is evident in its expropriations and lack of competitiveness. They keep refusing to adapt to a market model that would benefit their society and would enable them to develop more rapidly. The result is that they are not moving up. The Venezuelan government cannot ensure that the same laws will be in force next month. They are not inventing: there is no room to be creative when the incentives are not right.

India, although considered an emerging power, is a country where economic inequality has a profound knock-on effect. Child labour, insecurity, discrimination, gender inequality, and lack of health and education (mainly in rural areas) are not helping this country. They have recast their laws, but they are failing to enforce them. The police force is so corrupt that people are sometimes more afraid of going to the police to report a crime than of the crime itself.

With the second largest population in the world, India is a complex country with unsettling realities and many subcultures, which makes it a difficult country to handle. India ranks fairly low in the R^2 index, far below their BRIC peers. China has a life expectancy of more than seventy-five, whereas India's is seven years less, at almost sixty-eight. Fifty-five per cent of the Indian population is in multi-dimensional poverty; the Chinese figure is just under 6 per cent. In the gender arena, China is also doing better in any single field: it has more women holding seats in parliament, and almost twice as many who are educated than in India. Maternal mortality is alarming in India: 200 women die per 100,000 live births versus 37 in China. Businesses are starting to flourish in India and foreign investment is flowing, but due to bureaucracy and corruption it is still difficult to start a business.

Lack of clean water supplies and basic services are still a problem in countries like China, Indonesia, India, Kenya, Pakistan and Bangladesh. The governments of these countries are failing to provide not only shelter and water, but also education and health care for their populations. The survival and security results found in the index reflect this reality. For example, in April 2013 in Bangladesh, 1,129 people died when a building collapsed due to the lack of proper building regulations.

In the following pages you will find three charts showing countries ranked according to their R^2, C^2 and bio-logical values respectively. Look for your country and see how it's doing. If you're interested in learning more about this, there's a section in the Appendix that explains how we built the index.

In the Appendix, you will also find a table which ranks the

seventy-one countries according to the R^2 Mobility Index. The first column ranks the countries in terms of their R^2 scores from highest to lowest, while the second column indicates each country's R^2 value. The third column indicates its bio-logical value, which is composed of the average of the Four S's, and next to this column you can find the country ranking according to the bio-logical value alone. The fourth column tells us the value of the C^2, and on the right is the country ranking considering only the C^2.

The countries at the top of the table are those with the best combination of conditions to allow mobility, since they make a good use of their bio-logical make-up and culture codes for creating opportunities. The latter means that people use their Four S's to take the best opportunities that the economy and institutions provide in order for them to move up.

On the other hand, the countries at the bottom of the table are those whose governments do not provide social and economic conditions for mobility; or, in the cases where governments do provide these conditions, people do not make the best of their bio-logical and culture code opportunities for moving up. In some cases, a country with a high bio-logical value may have a low C^2 value, but what matters is the combination of the two.

It is also interesting to note that the countries at the top of our indicator reveal an important reality: size matters. Six out of the ten countries at the top of our R^2 Mobility Index have a population of less than 10 million: Switzerland, Singapore, Finland, Austria, Denmark and New Zealand. This is not a coincidence.

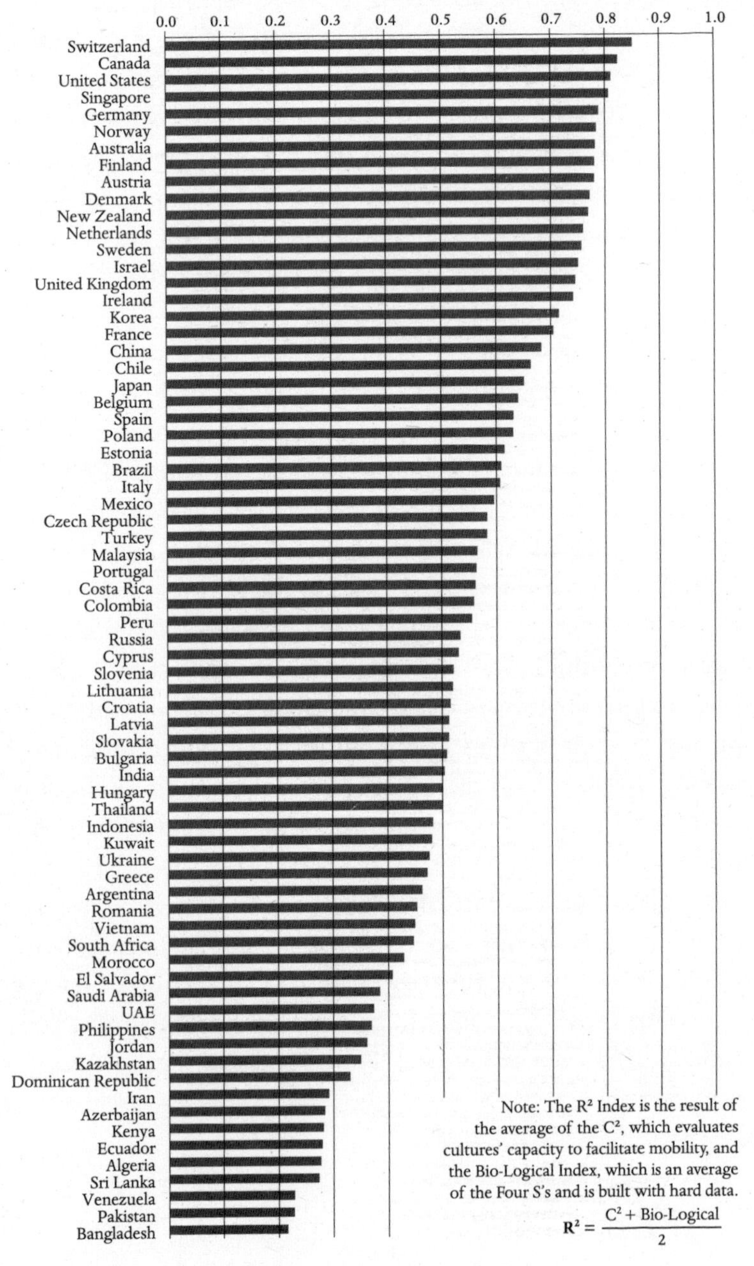

The R^2 Mobility Index

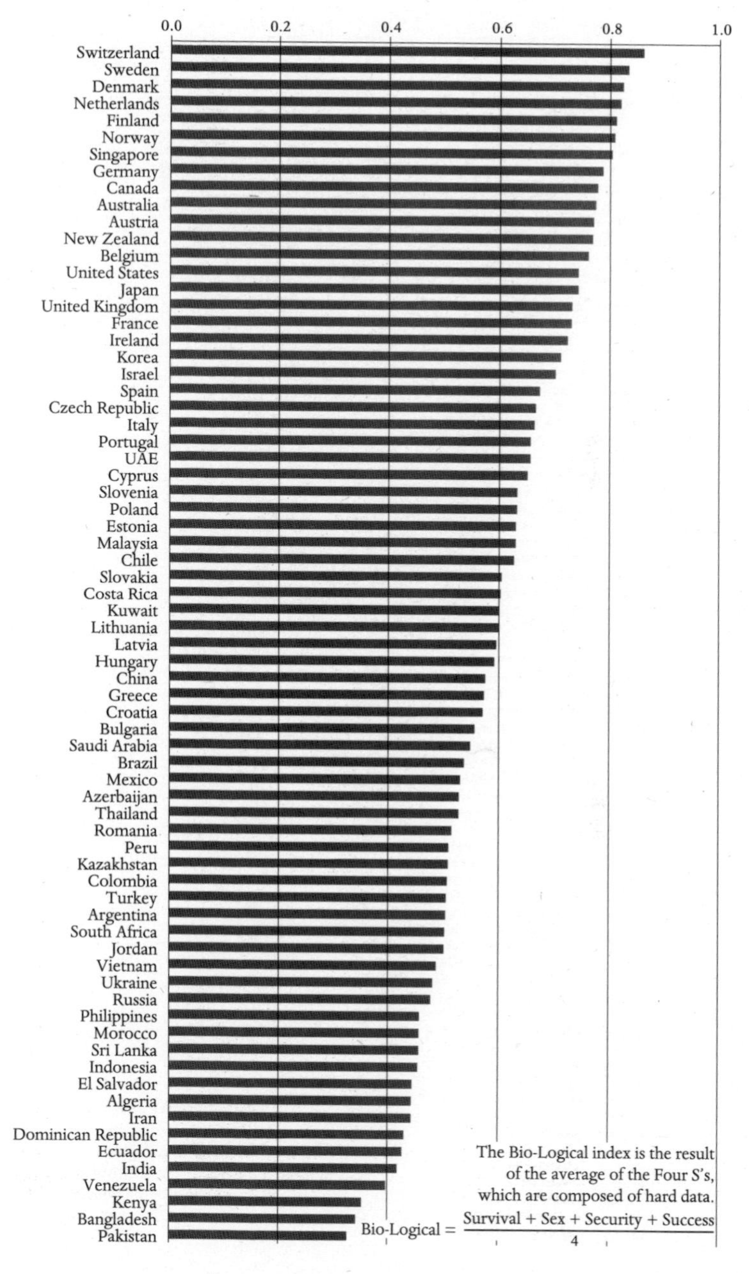

The Bio-Logical Index

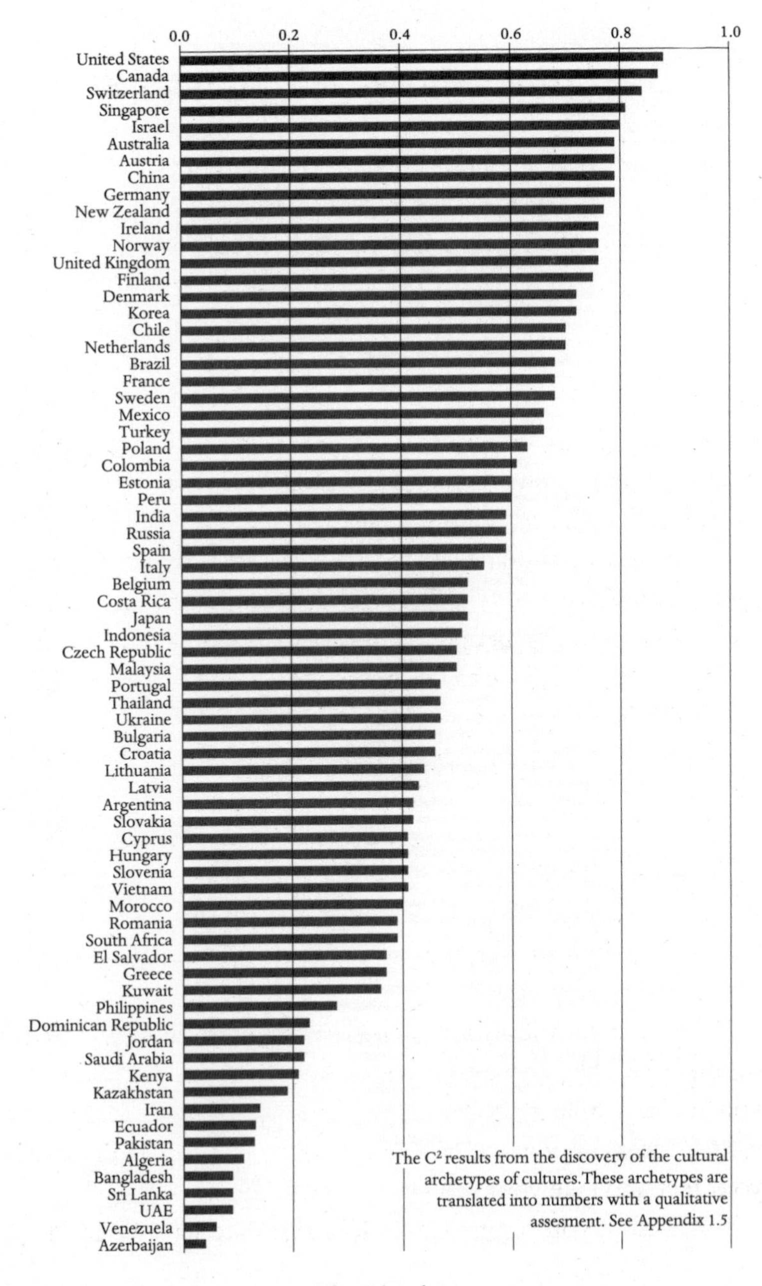
0.0
0.2
0.4
0.6
0.8
1.0
United States
Canada
Switzerland
Singapore
Israel
Australia
Austria
China
Germany
New Zealand
Ireland
Norway
United Kingdom
Finland
Denmark
Korea
Chile
Netherlands
Brazil
France
Sweden
Mexico
Turkey
Poland
Colombia
Estonia
Peru
India
Russia
Spain
Italy
Belgium
Costa Rica
Japan
Indonesia
Czech Republic
Malaysia
Portugal
Thailand
Ukraine
Bulgaria
Croatia
Lithuania
Latvia
Argentina
Slovakia
Cyprus
Hungary
Slovenia
Vietnam
Morocco
Romania
South Africa
El Salvador
Greece
Kuwait
Philippines
Dominican Republic
Jordan
Saudi Arabia
Kenya
Kazakhstan
Iran
Ecuador
Pakistan
Algeria
Bangladesh
Sri Lanka
UAE
Venezuela
Azerbaijan
The C^2 results from the discovery of the cultural archetypes of cultures.These archetypes are translated into numbers with a qualitative assesment. See Appendix 1.5

The C^2 Index

Conclusion: Voting With Your Feet

> You have brains in your head.
> You have feet in your shoes.
> You can steer yourself any direction you choose.
>
> Dr Seuss, *Oh, the Places You'll Go!*[1]

It's not about where you are, it's about where you're going. Your culture may be great, but if you're moving down or even staying still you're in trouble. France may have a great culture, but everybody is moving away so the culture is actually under threat. Colombia had a bad image for many years, but things are getting better so people are moving up. China may not be the most pleasant place to live but a lot of people are most definitely moving up there.

There's a belief that bumblebees should technically be incapable of flight, given their weight to wing power ratio. But because bumblebees are not aware of this reality they are able to fly anyway. So what should we do when our culture tells us we can't move up? We shouldn't believe everything our culture says. If your culture tells you not to move up, don't believe it, say no and fly anyway.

Why did the settlers move to America? Because they wanted to move up: they chose to leave, they voted with their feet. If you make a million dollars a year and don't want to spend most of your money on taxes in Sweden, you can vote with your feet and take your money to the Bahamas.

But, before you can make choices and vote with your feet,

you have to understand your priorities. It's not just about moving to other countries, it's also about transforming the culture code of your own country. That's why we developed the R^2 Mobility Index incorporating the C^2 and the Four S's. These are the variables that influence upward social mobility.

We cannot judge the whole of a culture from a single perspective, because we believe that no culture is solely good or bad, right or wrong. Just because one aspect of a culture may not be optimal for moving up doesn't mean that we have to throw out the baby with the bathwater, giving up our idiosyncrasies, customs and rituals. Rather we must be cautious of those reversal or crystallization laws that impede our movement. This is why our index analyses a culture in parts, explaining which features enable people to move up and which don't.

You might value security and move to Switzerland, but then you might find that life is boring there. You might want to live in Berlin because it's fun, but it's too big for your taste. You could review all the cultures we present in the index and focus on what each has to offer you, then take a look at your culture: is it good for you or is it bad for you? If you live in an Eskimo community and you want to have five husbands, that's fantastic, but the trade-off is that you're going to have to cook for those five husbands three times a day every day!

What we are trying to say is that there is a trade-off. You have to give up some benefits for others. Nothing in life is perfect, and there is no perfect culture. I may love Paris because I love the architecture and the arts, but the trade-off is that working in Paris can be unpleasant, expensive and the Parisian attitude may not suit my outlook. I want the comforts of social security in the Nordic countries, but the trade-off is months of darkness and unbearable cold. I may have fun in Tijuana, but the trade-off is that I need to have a bulletproof car. But in order to make the best choice you need to be aware of your options, and that is what the R^2 Mobility Index is for.

Having said that, naturally there are some cultures that offer greater opportunities for everyone to climb the social ladder and move up than others. If a culture wants its society to move up, its myths, icons, heroes, heroines and symbols should always be telling you, 'Move up, young man or woman!'

Throughout history, we have seen that the most successful cultures are those that have been able to appreciate and preserve the best aspects of their cultural heritage while at the same time have been willing to sail the seas of innovation and seek new horizons. These are societies that are open to change and unafraid to challenge the status quo. They are highly creative cultures, as was England's Elizabethan era, nineteenth-century Paris or Weimar Berlin.

This book shows that there are some universal cultural indicators that induce a culture to move down and ultimately die, or to move up and evolve. For example, unconscious cultural archetypes that are resigned to the fact that the world of the supernatural is what determines one's survival, or cultures which consider pleasure to be a sin, where sexual life is strongly influenced by religion, where women's rights are not respected, or where conspiracy theories hold that destiny, globalization, the CIA, karma, God or the Devil are responsible for our well being, will inevitably move down.

Countries most likely to move up are those that have cultural archetypes that emphasize that our future does not depend on supernatural powers nor on the poor distribution of resources, rather it depends on our capacity to deepen our knowledge of the world, our capacity to think and to create. For these reasons we can understand why Europe was able to surpass the Arab world for a millennium (from 800 to 1800), which precisely points to a bio-logical reason: European cities were more free to grow without interference since they were less repressed by the demands of predator states.[2]

We have found that cultures in which your own personal achievements and talents are what define success will always be looking upwards. Out of all the variables we analysed, we cannot emphasize enough that the most important variable is education. It is the fundamental condition for moving up. An education system that is dynamic and encourages creativity will stimulate competition. In a culture where this happens, where individuals can create or do things that nobody has ever thought of before, everyone falls into a kind of arena where those who strive for better things will reach the top. Of course, the battle should be fair, but that's where Survival, Sex and Security come into play: enforcing laws that promote individual freedom and gender equality, and fostering policies that ensure everyone's chances of evolving.

But the game of moving up is not just about cultural awareness; it's also about personal awareness. One of the key elements we study in neuroscience is the tension between the reality principle and the pleasure principle. The pleasure principle dictates that I love eating a lot of food, but reality says I might become obese, so I need to limit my intake in order to keep moving. That's the reality principle: you have to deal with the consequences. Trade-offs, and personal and cultural awareness, are guides that help us figure out what our options are for moving up in life.

We know that cultures matter and that they have a deep impact on us. We are born in a certain culture, and because of the imprints we have people's behaviour varies across cultures. An American, a Chinese, a Japanese and an Australian will not necessarily behave in the same way in a given situation, even though their basic instincts remain the same.

So cultures are very different and we would like to keep it that way. But one thing remains: we are bio-logical beings, and cultures deal with this reality in very different ways. When it comes to our biology, there is one critical condition that all cultural paths should consider: individuals and cultures must respect the reptilian. We are humans; we have impulses that permanently reside within us. We have made a point of emphasizing the important role the reptilian plays within our lives, and thus within our cultures. The reptilian brain is a crucial part of human nature and we cannot deny it, just as we cannot deny that the Earth is round, or that humans cannot fly (so far).

When Megan slept with Bradley, her biology took over. She did not think about the negative consequences that her actions might have on her relationship with Richard. It was her instincts reacting, her reptilian brain, and this is because the reptilian always wins.

We truly believe that humanity's greatest error has been its failure to comprehend what motivates us to move up. If we had been aware of our emotions – our fears, sexual attraction, concern for status, need for appreciation, enjoyment of risk, desire for success, enjoyment of surprise, need to belong, jealousy, lust for power, envy, love of freedom, love or happiness – the history of mankind would have been completely different.

Does this mean that we are impotent, that we are slaves to our biology? Not at all. As we have stated before, biology preconditions us, but we also need to know that we are not merely a bag of genes. The challenge is to learn how to harmonize our instincts (the reptilian) with our emotions (the limbic) and our logic (the cortex). Cultures can either embrace or punish our biology, and this in turn will have consequences for our

movement. What we propose in this book is a way of understanding our biology so that our cultures make the best of it and even celebrate it. This is how people and cultures move up.

One thing is clear: we have no choice – we have to keep moving, and moving up. Life is movement: to not move, to stop or to stand still is the worst thing one can do to our neurons and, in turn, to ourselves and our cultural context. As Albert Einstein once said, 'Life is like riding a bicycle: to keep your balance you must keep moving.'

When we raised our glasses at Missy's birthday party five years ago in Davos, Switzerland, we toasted a new mobility index. We knew we already had many indexes, among them GDP and the Happy Planet Index, but we really wanted to measure those motivations and drives linked to mobility. We hope that this book will merit another toast.

Still, we don't want to stop there. If we could achieve only one more thing through this book, it would be to challenge your way of thinking. And if we could even take this one step further, and we discover that you're combining different elements of this book into your life – looking at the tables and what aspects of a culture are mobility killers or mobility accelerators, and starting to think of ways to combine the best elements of each culture – then we are headed in the direction of creating a better, more stimulating and rewarding world.

Let's try it, even if they say it's impossible. It's UP to you!

Notes

Introduction

1. See D. Moyo, *Dead Aid: Why aid is not working and how there is a better way for Africa* (New York: Farrar, Straus and Giroux, 2009).
2. Daron Acemoglu and James Robinson, *Why Nations Fail: The origins of power, prosperity, and poverty* (New York: Crown Publishers, 2012).
3. Clotaire Rapaille, PhD has been analysing cultures and discovering archetypes for over thirty years, applying his discoveries to marketing. He holds a PhD in Social Psychology from the Sorbonne, and is the CEO and founder of Archetype Discoveries Worldwide, a company that seeks to uncover cultural codes for his clients. Andrés Roemer has a PhD in Public Policy from UC Berkeley, Master of Public Administration from Harvard University, BS in economics from Instituto Tecnológico Autónomo de México, and a BA in Law from Universidad Nacional Autónoma de México. He has been analysing human behaviour for the last fifteen years, particularly in the field of evolutionary psychology, and has published more than fifteen books on this subject.
4. G. T. Allison, R. D. Blackwill and A.Wyne, *Lee Kuan Yew: The grand master's insights on China, the United States, and the World* (Cambridge, MA: MIT Press, 2013).
5. Not only is it important for governments and cultures to help people move up, but every member of society plays an important role. One clear example is Ricardo Salinas Pliego's

'Limpiemos Nuestro México' (Let's clean our Mexico) proposal, which encourages people to be aware of environmental destruction and the waste we produce, and to do something about it in order to have a cleaner environment in Mexico. See also George L. Kelling and James Q. Wilson, 'Broken Windows', *The Atlantic*, March 1982; http://www.theatlantic.com/magazine/archive/1982/03/broken-windows/304465/ for another example. As Pliego says, 'If a window in a building is broken and is left unrepaired, all the rest of the windows will soon be broken.'

6. D. Kahneman, *Thinking, Fast and Slow* (New York: Farrar, Straus and Giroux, 2011).
7. M. Weber, *The Protestant Ethic and the Spirit of Capitalism* (London: Routledge, 2001).
8. N. Ferguson, *Civilization: The west and the rest* (New York: Penguin Books, 2012).
9. Clotaire Rapaille, *The Culture Code: An ingenious way to understand why people around the world live and buy as they do* (New York: Crown Business, 2006), p. 21.
10. H. Spencer, *Principles of Biology*, (London: Williams and Norgate, 1864) vol. 1, p. 444.

1 Back to Basics

1. A. Schopenhauer, *Essays on Freedom of the Will* (New Jersey: Wiley-Blackwell, 1995).
2. For those interested in how sex and desire influence our actions, check out Matt Ridley, *The Red Queen: Sex and the evolution of human nature* (London: Viking, 1993), which analyses the dichotomy between monogamy and polygamy; or D. H. Hamer and P. Copeland, *Living With Our Genes: Why they matter more than you think* (New York: Anchor, 1999), which discusses the importance

of human genes in our behaviours and actions. And see in particular Terry Burnham and Jay Phelan, *Mean Genes: From sex to money to food, taming our primal instincts* (London: Simon & Schuster, 2000) and Richard Dawkins, *The Selfish Gene* (Oxford: Oxford University Press, 1976), and find out how our genes guide us throughout our lifetime, especially when it comes to sex. For an explanation of how consumerism influences the way we search for a mate, see G. Miller, *Spent: Sex, evolution, and consumer behavior* (New York: Viking, 2009).

3. According to Alain de Botton's book *How to Think More About Sex* (New York: The School of Life, 2012), p. 53, 'Symmetry and balance matter so much to us because their opposites – facial asymmetry and imbalance – are markers of diseases contracted either in the womb or in the early years of life, at a time when the greater part of the self is still being shaped.' This is important to us because the main feature we look for when it comes to beauty is health: we want someone who both appears to be and actually is healthy in order to increase our chances of reproduction and survival. See also N. L. Etcoff, *Survival of the Prettiest: The science of beauty* (Michigan: Anchor, 2000).
4. F. C. Fang and A. Casadevall, 'Why We Cheat', *Scientific American Mind*, May/June 2013.
5. It's not a coincidence that the most famous stories throughout history have told tales of romance, cheating and jealousy. Want to find out more? Check out D. P. Barash and N. R. Barash, *Madame Bovary's Ovaries: A Darwinian look at literature* (New York: Delacourt Press, 2005). For cinema lovers there are three perfect examples of why stories of sex and cheating are always so attractive. First, *Quartet*, directed by Dustin Hoffman, specifically the scene where a girl recalls how she slept with a great tenor just before she got married and then, as was the case with Megan, felt the need to confess. *Hyde Park on Hudson*, directed

by Roger Michell, describes the love affair between the former President of the United States Franklin D. Roosevelt and a distant cousin called Margaret Suckley. The final example is *Closer*, a great movie directed by Mike Nichols in which we see four people building and destroying all the loving relationships an individual can develop.

6. For a complete, in-depth account of how Deng Xiaoping established China's economic development during the 1980s and how this challenged the conventional wisdom of Western economic theory and policy, see Ezra F. Vogel, *Deng Xiaoping and the Transformation of China* (Cambridge, MA: Belknap Press, 2011).
7. See J. Brockman, *This Explains Everything: Deep, beautiful, and elegant theories of how the world works* (New York: Harper Perennial, 2013).

2 *The Reptilian Always Wins*

1. New York: Dover Publications, 2000.
2. Dan Ariely explains that there's a correlation between intelligence, creativity and cheating. According to his studies, if a person is more creative, then the probabilities of cheating are greater. In this sense we can argue that one of the reasons why Clinton tried to fool the nation was because he trusted his intelligence to get away with it. For a detailed explanation, see http://www.scientificamerican.com/podcast/episode.cfm?id=dan-ariely-talks-creativity-and-dis-12-12-29

   Another explanation could be that although rational theory tells us that our behaviour is all about cost v. benefit, the reality is that we don't actually think in terms of long-term consequences as much as we think we do. One of the main influences is seeing other people getting away with it. Dan Ariely (2008)

Predictably Irrational: The hidden forces that shape our decisions (New York: HarperCollins Publishers).

3. Paul D. MacLean, *The Triune Brain in Evolution: Role in paleocerebral functions* (New York: Plenum Press, 1990).
4. Found in every species of animal, the reptilian guides our most basic instincts. As humans, we tend to think of ourselves as separate from the animal kingdom, blessed with rationality and intellect, but we are still animals. On this topic, Jared Diamond's *The Third Chimpanzee: The evolution and future of the human animal* (New York: HarperCollins, 1992) and *The Rise and Fall of the Third Chimpanzee: How our animal heritage affects the way we live* (New York: Vintage Books, 2003) are worth a read.
5. D. M. Buss, *The Evolution of Desire: Strategies of human mating* (New York: Basic Books, 2003).
6. We all lie. Sometimes our lies are big, sometimes they are small, sometimes we lie about our lies, and sometimes we even believe our own lies. David Livingstone Smith's book *Why We Lie: The evolutionary roots of deception and the unconscious mind* (New York: St Martin's Press, 2004) is an interesting read if you want to delve deeper into the world of deception.
7. For more information about how soccer can be understood as a reptilian activity, see A. Roemer and E. Ghersi, *¿Por qué amamos el futbol? : Un enfoque de derecho y economía* (Mexico: Porrua, 2008).
8. Though polemic, an interesting read on sperm and how they impact human behaviour is Robin Baker, *Sperm Wars: Infidelity, sexual conflict, and other bedroom battles* (New York: Basic Books, 2006). Robin Baker and Mark Bellis, *Human Sperm Competition: Copulation, masturbation and infidelity* (London: Chapman & Hall, 1995) details the effects of sperm competition on human sexual behaviour and fertility issues.
9. David Buss and Cindy Meston mentioned during their presentation at the University of Texas at Austin that, in one of their

surveys, women described 237 reasons why they had sex. These reasons ranged from 'the very altruistic, of having sex because they wanted to please their partner, to the vengeful, of getting back at a partner who cheated on them by having sex with (the ultimate revenge) their best friend, to the borderline evil of some women who said that they would intentionally have sex to try to transmit a sexually transmitted infection'. Motives ranged from adventure, exploration and experience to conquest – in which women compete with each other; or from the very mundane to the medicinal. From duty to pleasure, or bonding to pure physical pleasure. David Buss and Cindy Meston at CASW 2009, *Why Do Women Have Sex?*; http://www.youtube.com/watch?v=KAosqg3EHm8 In 2010, at Ciudad de las Ideas, a forum for great minds and ideas, Dr Andrés Roemer asked David Buss, 'So if women have 237 reasons to have sex, what about men? How many reasons do they have?' to which David Buss immediately answered, 'Men have one reason: if they have the chance.'

10. For more on this, see Mark Layner and Billy Goldberg, *Why Do Men Fall Asleep After Sex? More questions you'd only ask a doctor after your third whiskey sour* (New York: Three Rivers Press, 2006).
11. For many centuries, the human orgasm was a scientific mystery. Modern science has come a long way, and the orgasm has become a hot topic. Find out more about orgasms in Wilhelm Reich, *The Function of the Orgasm* (New York: Farrar, Straus and Giroux, 1973).
12. To better understand the connection between morals and how we see the sexual act, check out Alain de Botton, *How to Think More About Sex* (New York: The School of Life, 2012).
13. For more information, see R. Pool, *Eve's Rib: Searching for the biological roots of sex differences* (New York: Crown Publishers, 1994), p. 12.

14. The Natalie Wood quote is from N. M. Henley and J. D. Goodchilds, *The Curmudgeon Woman* (Missouri: Andrews McMeel Publishing, 2000).
15. Why are we monogamous? Why is divorce still so prevalent? How do other cultures perceive sex? These are just a few of the many questions Christopher Ryan and Cacilda Jethá touch on in their book *Sex at Dawn: How we mate, why we stray, and what it means for modern relationships* (New York: HarperCollins, 2010). For more about the mating game, see D. P. Barash and N. R. Barash, *Madame Bovary's Ovaries: A Darwinian look at literature* (New York: Delacourt Press, 2005).
16. Tom Standage, *A History of the World in 6 Glasses* (USA: Walker Publishing Company, 2006).
17. Pleasure is strongly linked to the reptilian. As Paul Bloom puts it in *How Pleasure Works: The New Science of Why We Like What We Like* (New York: W.W. Norton & Company, 2010), pleasure is not an evolutionary adaptation, but rather a by-product: accidents that help us survive and reproduce. This explains the huge variety in pleasure different people take in experiences, from sexual fetishes that involve vomit or faeces, to the pleasure some individuals experience when meeting a celebrity. Bloom uses the term 'essentialism', referring to those reptilian things that have a true nature or essence; but it's the hidden nature of these essentials that interest us, that give rise to different sources of pleasure.
18. October 2010, 'Norway petroleum fund tops $500bn' (BBC); http://www.bbc.co.uk/news/business-11571650

3 Time, Space and Energy

1. We highly recommend all of his books, including *The Silent Language, The Hidden Dimension, The Dance of Life, Beyond Culture* and *Understanding Cultural Differences: Germans, French and Americans*; see Bibliography for full details.
2. But, like our hunter-gatherer ancestors, we also need patience and a strategy. Friends don't just appear out of nowhere and start praising you; it's important to hone the relationships and feed their egos as well. Find out more in Frank Rose, 'The Selfish Meme' in *The Atlantic*; http://www.theatlantic.com/magazine/archive/2012/10/the-selfish-meme/309080/
3. The psychologist Walter Mischel has been conducting research in social psychology for over fifty years in universities that include Harvard, Stanford and Columbia. His famous marshmallow experiment has been replicated and altered by many psychologists since it was first carried out. W. Mischel, Y. Shoda and M. L. Rodriguez, 'Delay of Gratification in Children', *Science*, 244, pp. 933–8.
4. In his book *Thinking, Fast and Slow* (New York: Farrar, Straus and Giroux, 2011), the Nobel Prizewinner in Economics, Daniel Kahneman, divides the way brain forms thoughts into two categories: system 1, which is completely reptilian, is an automatic and unconscious process; system 2, which is the limbic and cortex, is a conscious, calculated and slow effort.
5. Our perception of social and personal space has a greater influence than just over our physical space. According to Edward T. Hall we inhabit different sensory worlds. The architectural environments that we create are expressions of our filtering process. Men and women inhabit different visual worlds; they have learned to use their eyes in different ways. This selective screen-

ing of sensory data allows some of the stimuli through whilst filtering out things that are not part of our cultural pattern. The term coined by Hall, 'Proxemics', is given to the interrelated theories and observations of the use of space as a specialized elaboration of culture. Edward T. Hall, *The Hidden Dimension* (New York: Anchor Books, 1966).

6. If you're interested in an alternative theory about how we can interpret religious art, see Alain de Botton, *Religion for Atheists: A non-believer's guide to the uses of religion* (New York: Vintage, 2013).
7. The 'just do it' culture has many side effects and consequences for society, but one interesting one is that it makes people less objective. According to psychologists William Hart and Dolores Albarracin, the 'action-first' attitude encourages them to act first out of their preconceived notions, as opposed to receiving information from the outside that differs from their own in order to possibly make better choices. See Tom Jacobs, '"Just Do It!" Culture Feeds Confirmation Bias', *Pacific Standard*, 25 May 2011; http://www.psmag.com/culture-society/just-do-it-culture-feeds-confirmation-bias-31496/

4 The Ideal Scenario

1. In his book *The Human Zoo* (New York: Kodansha International, 1996), Desmond Morris discusses how people living in societies, and especially those living in cities, resemble life in a zoo, a human zoo. With all their needs for survival (food and shelter) covered in most cities, humans still have trouble with relationships, with work and with the city's structure itself. See especially Chapter 3, 'Sex and Super Sex', where he talks about how some buildings resemble the image of the penis as a display of power.

2. In *The Great Convergence* (New York: PublicAffairs, 2013), Kishore Mahbubani discusses how, facing an inevitably globalized and interdependent world in which the western countries might not rule in a near future, the best thing we can do is to strengthen international institutions. This is particularly important for avoiding international conflicts, just as Europe has managed to do through the European Union.
3. Karma is not the only crystallized idea that people have. A westernized version of Karma is the law of attraction, which states that you get what you think of. Think of disease and you'll get disease. Even though there's no scientific evidence for karma or the law of attraction, these ideas have a profound impact on society and demonstrate how a myth can affect the behaviour of some individuals. For example, by the year 2010, new age bestseller *The Secret* had sold a reported 19 million copies and been translated into forty-five languages (as reported in 'Coming Soon: Sequel of *The Secret*', *Hindustan Times*, 16 July 2010; http://www.hindustantimes.com/News-Feed/Books/Coming-soon-Sequel-of-The-Secret/Article1-573262.aspx).
4. A good example of how crystallization works is the horoscope: if we allow our reptilian instinct to believe in destiny rather than use our cortex to explain why things are happening, our chance of moving up is lower. For more about horoscopes and their influence on people, see R. Pool, *Eve's Rib: Searching for the biological roots of sex differences* (New York: Crown Publishers, 1994).
5. From *The Myth of Sisyphus, and Other Essays* (New York: Vintage, 1991).
6. http://www.starbucks.co.uk/menu/beverage-list/espresso-beverages

5 The Five Critical Moves

1. This section is based on the methodology developed by Dr Clotaire Rapaille. For a detailed explanation of this, see his book *The Culture Code: An ingenious way to understand why people around the world live and buy as they do* (New York: Crown Business, 2006)
2. Ibid., p. 21.
3. William James, *The Principles of Psychology* (New York: Dover Publications, 1950).
4. Antti Revonsuo believes that modern humans experience the same dreams as our ancient ancestors, because dreams seem to be biologically programmed into our brain. Moreover, nightmares are rehearsals and expressions of our instincts in order to overcome danger. See his *Inner Presence: Consciousness as a biological phenomenon* (Cambridge, MA: MIT Press, 2006).
5. Donald Brown says, 'Human universals – of which hundreds have been identified – consist of those features of culture, society, language, behavior, and mind that, so far as the record has been examined, are found among all peoples known to ethnography and history.' Donald Brown, *Human Universals* (New York: McGraw-Hill, 1991).
6. For a detailed explanation about the non-existence of the 'blank slate' and its implications for human behaviour, check out Steven Pinker, *The Blank Slate: The modern denial of human nature* (New York: Penguin Books, 2003).

6 The Third Unconscious

1. C. J. Emden, *Nietzsche on Language, Consciousness, and the Body* (Champaign: University of Illinois Press, 2005).

2. For an introduction to the field of evolutionary psychology, see D. M. Buss, *Evolutionary Psychology: The new science of the mind* (Boston: Allyn and Bacon, 2007) and *The Handbook of Evolutionary Psychology* (Hoboken: John Wiley, 2005); Steven Pinker, *How the Mind Works* (New York: W. W. Norton, 2009); J. H. Barkow, L. Cosmides and J. Tooby, *The Adapted Mind: Evolutionary psychology and the generation of culture* (Oxford: Oxford University Press, 1995); E. O. Wilson, *Sociobiology: The new synthesis* (Cambridge, MA: Belknap Press, 2000); L. Barret and R. Dunbar, *The Oxford Handbook of Evolutionary Psychology* (Oxford: Oxford University Press, 2009); R. Wright, *The Moral Animal: Why we are the way we are – the new science of evolutionary psychology* (New York: Vintage, 1995). Also, J. C. Confer et al., 'Evolutionary Psychology: Controversies, questions, prospects, and limitations' *American Psychologist*, 65 (2), pp. 110–26; D. P. Schmitt and J. J. Pilcher, 'Evaluating Evidence of Psychological Adaptation: How do we know one when we see one?', *Psychological Science*, 15, p. 643 (available at http://www.bradley.edu/academic/departments/psychology/faculty/schmitt/).
3. Freud developed the first theory of the unconscious through the interpretation of dreams in his book *The Interpretation of Dreams* (1899). He believed that dreams were the channel for our unconscious to communicate with our conscious mind to tell us something about our behaviour.
4. Jung's theory about archetypes describes the unconscious as having a collective level where we share similar symbols or images that control some actions. In the chapter 'Approaching the Unconscious', he argues that mankind shares a common memory that is always represented in our dreams, where symbols or images are projected as an 'individual' reaction but their interpretation is collective and therefore archetypes are created from there. (Carl Jung, *Man and His Symbols*; New York: Dell, 1968.)

5. Another example of a personality test based on archetypes is the Myers–Briggs Type Indicator developed by Katharine Cook Briggs and her daughter Isabel Briggs Myers as a way of determining personality types, or archetypes, in order to function as a guide for individuals who want to enter the labour force but don't know what work they might be good at. See Isabel Briggs Myers with Peter B. Myers, *Gifts Differing: Understanding personality type* (Boston: Nicholas Brealey, 1995).
6. See Carl Jung and Anthony Storr (ed.), *The Essential Jung: Selected writings* (Princeton: Princeton University Press, 1999).
7. Clotaire Rapaille, *The Culture Code: An ingenious way to understand why people around the world live and buy as they do*. New York: Crown Business, 2006), p. 5.
8. Laborit demonstrated that knowledge is more likely to remain in our memory if there is a strong experience attached to the process of learning. So if a person lives the experience they are trying to learn, it becomes easier for them to fully comprehend the concept. With this in mind, he developed the theory that an experience attached to learning generates what he calls an imprint. Erich Fromm also said that ideas and thoughts only influence man deeply when they are taught through a living experience, and that merely to 'become acquainted with other ideas is not enough, even though these ideas in themselves are right and potent. But ideas do have an effect on man if the idea is lived by the one who teaches it; if it is personified by the teacher, if the idea appears in the flesh. If a man expresses the idea of humility and is humble, then those who listen to him will understand what humility is. They will not only understand, but they will believe that he is talking about reality.' *On Disobedience and Other Essays* (New York: Seabury Press, 1981).
9. Alain de Botton, *How to Think More About Sex* (New York: The School of Life, 2012).

7 The Four S's

1. Maslow developed his theory of needs in a famous article published in 1943: Abraham H. Maslow, 'A Theory of Human Motivation', *Psychological Review*, 50, pp. 370–96.

8 Survival

1. Ezra F. Vogel, *Deng Xiaoping and the Transformation of China* (Cambridge, MA: Belknap Press, 2011), p. 41.
2. Ibid., p. 54.
3. Compared to other species in the animal kingdom, humans take the longest to develop: the development process takes over a decade until the individual can become completely independent. This seems like an evolutionary puzzle since there are high costs attached to development. To find out why, check out David F. Bjorklund (2007) *Why Youth Is Not Wasted on the Young: Immaturity in human development* (Hoboken: Wiley-Blackwell).
4. A family is not simply the nuclear ideal of two parents and children. It's much more complex than that. It's about communication, signalling, rearing, indoctrination and rites of passage. It's worth taking a look at Desmond Morris, *The Human Animal: A personal view of the human species* (London: BBC Books, 1994).
5. R. A. Spitz, 'Hospitalism: An inquiry into the genesis of psychiatric conditions in early childhood', *Psychoanalytic Study of the Child*, 1, pp. 53–74.
6. For a complete and accurate study about how religions are perceived by atheists, see Alain de Botton, *How to Think More About Sex* (New York: The School of Life, 2012).

7. Subir Bhaumik, '"Witch" Family Killed in India', BBC *News*, Calcutta, 12 June 2008; http://news.bbc.co.uk/2/hi/south_asia/7449825.stm
8. Kundan Pandey, 'Whole of India Needs Anti-Superstition Law', *Down to Earth*, 22 August 2013; http://www.downtoearth.org.in/content/whole-india-needs-anti-superstition-law
9. 'Narendra Dabholkar', *The Economist*, 14 September 2013; http://www.economist.com/news/obituary/21586275-narendra-dabholkar-fighter-against-superstition-was-killed-august-20th-aged-67-narendra
10. 'New School Values', *The Economist*, 11 September 2014; http://www.economist.com/news/international/21616978-higher-teacher-pay-and-smaller-classes-are-not-best-education-policies-new-school
11. Adam Zamoyski, *Moscow 1812: Napoleon's fatal march* (New York: Harper Perennial, 2005).
12. B. F. Skinner, '"Superstition" in the Pigeon', *Journal of Experimental Psychology*, 38, pp. 168–72 (1947).

9 Sex

1. According to Robert Pool (*Eve's Rib: Searching for the biological roots of sex differences*; New York: Crown Publishers, 1994), there is a growing amount of evidence regarding the influence that sex hormones have over most of our behaviour. In other words, they don't only have an impact on our emotional behaviour, but also on our cognitive abilities.
2. Evolutionary psychology provides great explanations for why we enjoy sex, and why we feel the need to have sex. However, we are much more than just the pure result of an evolutionary process. Just as Alain de Botton says in *How to Think More About Sex* (New York: The School of Life, 2012), these explanations,

albeit true, are in the end boring and apparently incomplete. We do enjoy sex because we evolved this way, otherwise we would not have reproduced as we have; but evolution does not explain what we are thinking when we get nervous as we talk to the woman or man we like, or how we feel as we undress ready to have sex. In de Botton's words, 'sex will never be either simple or *nice* in the ways we might like it to be. It is not fundamentally democratic or kind; it is bound up with cruelty, transgression and the desire for subjugation and humiliation.' When we have sex, we return to that moment in childhood when we were loved and embraced. We overcome loneliness, and become intimate with our partner. Having sex for no other reason than to simply fornicate, without any bonding, produces the same effect in the brain as love-making, but does not produce the same effect in our feelings, in the limbic.

3. Jared Diamond, *Why Is Sex Fun? The evolution of human sexuality* (New York: Perseus Books, 1997).
4. Ibid., p. 1.
5. 'More Bang for your buck', *The Economist*, 9 August 2014; http://www.economist.com/news/briefing/21611074-how-new-technology-shaking-up-oldest-business-more-bang-your-buck
6. 'Throwing off the Covers', *The Economist*, 7 August 2014; http://www.economist.com/news/middle-east-and-africa/21611117-official-report-blows-lid-secret-world-sex-throwing.
7. http://gingerrogers.com/about/quotes.html
8. John Medina (*Brain Rules: 12 principles for surviving and thriving at work, home and school*; New York: Pear Press, 2009) believes that men and women process certain emotions differently and that these differences are a by-product of the complex interactions between nature (our genetic information) and nurture (our culture and its paradigms transmitted to us through language).

9. In February 2013 the *Harvard Business Review* wrote about how female leaders should handle double standards: 'The premise is that women have not been socialized to compete successfully in the world of men, and so they must be taught the skills their male counterparts have acquired naturally. But, at the same time, they must "tone it down" or risk being labeled as having sharp elbows.' In other words it is a 'damned if you do, damned if you don't' dilemma. Women are constantly evaluated against masculine standards, and the actual work that they do and the results they produce are secondary in many cases. http://blogs.hbr.org/cs/2013/02/how_female_leaders_should_handle_double_standards.html
10. John Medina also suggests that we could have environments where gender differences are both noted and celebrated instead of being ignored or marginalized. Every environment needs to take advantage of the qualities that both genders bring to the table. *Brain Rules*, p. 259.
11. Donna Abu-Nasr, 'Rape Case Roils Saudi Legal System', *Washington Post*, 21 November 2006; http://www.washingtonpost.com/wp-dyn/content/article/2006/11/21/AR2006112100967.html
12. Max Fisher, 'Saudi Arabia's Oppression of Women Goes Way Beyond its Ban on Driving', *Washington Post*, 28 October 2013; http://www.washingtonpost.com/blogs/worldviews/wp/2013/10/28/saudi-arabias-oppression-of-women-goes-way-beyond-its-ban-on-driving/
13. Raghabendra Chattopadhyay and Esther Duflo, 'Women as Policy Makers: Evidence from a randomized policy experiment in India', *Econometrica*, 72 (5), pp. 1409–43.
14. Amanda Fortini, 'The Feminist Reawakening', *New York Magazine*, 13 April 2008; http://nymag.com/news/features/46011/

10 Security

1. Andrew Jacobs, 'Too Old and Frail to Re-educate? Not in China', *New York Times*, 20 August 2008 (http://www.nytimes.com/2008/08/21/sports/olympics/21protest.html?pagewanted=all&_r=0); Zhu Keliang and Roy Prosterman, 'Land Reform Efforts in China', *China Business Review*, 1 October 2012 (http://www.chinabusinessreview.com/land-reform-efforts-in-china/).
2. R. A. Pastor, *Limits to Friendship: The United States and Mexico* (New York: Vintage, 2011).
3. Barry Buzan and Lene Hansen, *The Evolution of International Security Studies* (New York: Cambridge University Press, 2009).
4. Steven Pinker argues in his book *The Better Angels of Our Nature: Why violence has declined* (New York: Viking, 2011) how and why violence has declined over the years, even though this may not seem apparent to us.
5. Eduard Punset, *Viaje al optimismo: Las claves del futuro* (Mexico City: Editorial Diana, 2013), p. 200 (our translation).
6. Bruce Schneier recognizes that doing something about airport security is better than doing nothing, especially if the things you do are visible, such as airport security checks. However, when politics comes into play, the best solutions – investment in intelligence, for example – are not necessarily implemented because they are likely to be less visible, even though they are more effective. See his blog: http://www.schneier.com/blog/archives/2013/05/the_politics_of_3.html
7. David Montero, 'Bangladesh: The blowback of corruption' PBS, 21 August 2009; http://www.pbs.org/frontlineworld/stories/bribe/2009/08/bangladesh-a-dirty-deal-back-fires.html

8. 'Legalize freedom' is the term that Robert Cooter and Hans-Bernd Schäfer use to refer to creating trust in institutions that can ensure the launch of innovative business ventures. 'Freedom is the presence of good law, not absence of all law,' they say. Law should be dependable and effectively designed to enhance not only personal security, but also growth and development. To read more about how law can change the world, see Robert D. Cooter and Hans-Bernd Schäfer, *Solomon's Knot: How law can end the poverty of nations* (Princeton: Princeton University Press, 2012).
9. Jack Perkowski, 'Protecting Intellectual Property Rights in China', *Forbes*, 18April 2012; http://www.forbes.com/sites/jackperkowski/2012/04/18/protecting-intellectual-property-rights-in-china/
10. International Centre for Prison Studies; retrieved 30 September 2014.
11. *The War on Marijuana in Black and White*, American Civil Liberties Union, June 2013.
12. *Report of The Sentencing Project to the United Nations Human Rights Committee Regarding Racial Disparities in the United States Criminal Justice System*, The Sentencing Project, August 2013.
13. Richard D. Knabb, Jamie R. Rhome and Daniel P. Brown for the National Hurricane Center, *Tropical Cyclone Report, Hurricane Katrina, 23–30 August 2005*, first published 20 December 2005; PDF available at http://www.nhc.noaa.gov/pdf/TCR-AL122005_Katrina.pdf

11 Success

1. Robert H. Frank, *Choosing the Right Pond: Human behavior and the quest for status*. (New York: Oxford University Press, 1985).

2. Y. Kim and Yinlong Zhang, 'The Impact of Power–Distance Belief on Consumers' Preference for Status Brands', *Journal of Global Marketing*, 27 (1), pp. 13–29.
3. Alain de Botton, *Status Anxiety* (New York: Pantheon Books, 2004). Also, for an interesting read, see R. Frank, *Richistan: A journey through the American wealth boom and the lives of the new rich* (New York: Three Rivers Press, 2007).
4. If you are interested in how cultures have been classified according to gender, look at the thorough study carried out by G. Hofstede, G. J. Hofstede and M. Minkov, *Cultures and Organizations: Software of the mind, international cooperation and its importance for survival*. (New York: McGraw-Hill, 2010), and G. Hofstede, G. J. Hofstede and P. B. Pedersen, *Exploring Culture: Exercises, stories and synthetic cultures* (Boston: Intercultural Press, 2002). For more about the cultural dimensions of Brazil or any other country, go to http://geert-hofstede.com/national-culture.html
5. To learn more about the term 'capability', check out Martha C. Nussbaum, *Creating Capabilities: The human development approach*. (Cambridge, MA: Harvard University Press, 2011).
6. B. L. De Mente, NTC*'s Dictionary of Mexican Cultural Code Words: The complete guide to key words that express how the Mexicans think, communicate, and behave* (New York: McGraw-Hill: 1996).
7. The British writer E. M. Forster in his essay 'Notes on the English Character' makes a wonderful comparison between English and French personalities: 'Once upon a time a coach, containing some Englishmen and some Frenchmen, was driving over the Alps. The horses ran away, and as they were dashing across a bridge, the coach caught on the stonework, tottered, and nearly fell into the ravine below. The Frenchmen were frantic with

terror: they screamed and gesticulated and flung themselves about, as Frenchmen would. The Englishmen sat quite calm. An hour later, the coach drew up at an inn to change horses, and by that time the situations were exactly reversed. The Frenchmen had forgotten all about the danger, and were chattering gaily; the Englishmen had just begun to feel it, and one had a nervous breakdown and was obliged to go to bed . . . The Frenchmen responded at once; the Englishmen responded in time. They were slow and they were also practical.' From *Abinger Harvest* (New York: Mariner Books, 1964)

8. See Thorsten Veblen, *The Theory of the Leisure Class: An economic study of institutions*, (Delhi: Aakar Books, 2005).
9. For a brief review of brands and their role in human behaviour, check out Matthew Hutson, 'Status Anxiety', *The Atlantic*, 17 September 2014; http://www.theatlantic.com/magazine/archive/2014/10/status-anxiety/379339/ For the academic references, see Nelissen and Meijers, 'Social Benefits of Luxury Brands as Costly Signals of Wealth and Status', *Evolution and Human Behavior*, 32 (5), pp. 343–55; D. D. Rucker and A. D. Galinsky, 'Desire to Acquire: Powerlessness and compensatory consumption', *Journal of Consumer Research*, 35 (2), pp. 257–67; K. Charles, et al., 'Conspicuous Consumption and Race', *Quarterly Journal of Economics*, 124 (2), pp. 425–67; Gad Saad, and John G. Vongas, 'The Effect of Conspicuous Consumption on Men's Testosterone Levels', *Organizational Behavior and Human Decision Processes*, 110 (2), pp. 80–92; J. M. Sundie et al., 'Peacocks, Porsches, and Thorstein Veblen: Conspicuous consumption as a sexual signaling system', *Journal of Personality and Social Psychology*, 100 (4), p. 664; Y. Wang and V. Griskevicius, 'Conspicuous Consumption, Relationships, and Rivals: Women's luxury products as signals to other women', *Journal of Consumer Research*, 40 (5), pp. 834–54; Y.

J. Han, J. C. Nunes and X. Drèze, 'Signaling Status with Luxury Goods: The role of brand prominence', *Journal of Marketing*, 74 (4), pp. 15–30.

10. The term *schadenfreude*, or rather the behaviour related to this feeling, has been present throughout the history of mankind. A good example is Victor Hugo's poem 'Envy and Avarice', a story of how a person can be happy as a result of someone else's disgrace, which is available online, for example at http://www.readbookonline.net/readOnLine/11841/
11. Daron Acemoglu and James Robinson, *Why Nations Fail: The origins of power, prosperity, and poverty* (New York: Crown Publishers, 2012).
12. See Dan Senor and Saul Singer, *Start-Up Nation: The story of Israel's economic miracle* (New York: Twelve, 2011). The authors explain how Israel, despite being in the middle of the desert and surrounded by enemies, has managed to develop and move up rapidly. They find the answer in innovation and an entrepreneurial culture.
13. Robert D. Cooter and Aaron Eldin, 'Law and Growth Economics: A framework for research' *Working Paper Series, Berkeley Program in Law and Economics*, UC Berkeley, 13 January 2011.
14. George Bernard Shaw said, 'If you have an apple and I have an apple and we exchange apples then you and I will still each have one apple. But if you have an idea and I have an idea and we exchange these ideas, then each of us will have two ideas.' Paul Roemer shows how human capital is the driver of growth, and population growth alone is not enough, as previously believed by Solow, the father of growth theory. This is a strong argument for encouraging entrepreneurship. See Paul Roemer, 'Endogenous Technological Change, Part 2: The problem of development: A conference of the Institute for the Study of Free Enterprise Systems', *Journal of Political Economy*, 98 (5), pp. S71–S102. Solow,

the father of growth theory, believed that it was the accumulation of capital and the population growth rate that made economies grow. So demonstrating that human capital does matter is a major breakthrough. Entrepreneurs and investors will appreciate A. Boynton, B. Fischer. and W. Bole, *The Idea Hunter: How to find the best ideas and make them happen* (San Francisco: Jossey-Bass, 2011), an excellent book on how and where to discover potential business ideas.

15. Ezra F. Vogel, *Deng Xiaoping and the Transformation of China* (Cambridge, MA: Belknap Press, 2011), p. 456.
16. World Intellectual Property Organization – Economics and Statistics Division, *World Intellectual Property Indicators, Special Section: The international mobility of investors*, 2013, p. 35.
17. See Robert Asprey, *The Reign of Napoleon Bonaparte* (New York: Basic Books, 2001), and Steven Englund, *Napoleon: A political life* (Cambridge, MA: Harvard University Press, 2004).
18. 'Social Mobility in America: Repairing the rungs on the ladder', *The Economist*, 9 February 2013; http://www.economist.com/news/leaders/21571417-how-prevent-virtuous-meritocracy-entrenching-itself-top-repairing-rungs

Conclusion: Voting With Your Feet

1. New York: Random House, 1990.
2. Eduard Punset, *Viaje al optimismo: Las claves del futuro* (Mexico City: Editorial Diana, 2013), p. 117.

Bibliography

Acemoglu, D. and Robinson, J. (2012), *Why Nations Fail: The origins of power, prosperity, and poverty*. New York: Crown Publishers.

Aczel, A. (2001), *The Riddle of the Compass: The invention that changed the world*. San Diego: Harcourt.

Aczel, A. (2002), *Fermat's Last Theorem: The story of a riddle that confounded the world's greatest minds for 358 years*. London: Fourth Estate.

Addiego, F., Belzer, E. G., Comolli, J., Moger, W., Perry, J. D. and Whipple, B. (1981), 'Female Ejaculation: A case study', *Journal of Sex Research*, 17, pp. 13–21.

Aharon, I., Etcoff, N., Ariely, D., Chabris, C. F., O'Connor, E. and Breiter, H. C. (2001), 'Beautiful Faces Have Variable Reward Value: fMRI and behavioral evidence', *Neuron*, 32.

Allison, G. T., Blackwill, R. D. and Wyne, A. (2013), *Lee Kuan Yew: The grand master's insights on China, the United States, and the world*. Cambridge, MA: MIT Press.

Ariely, D. (2009), *Predictably Irrational: The hidden forces that shape our decisions*. New York: Harper Perennial.

Asprey, R. (2001), *The Reign of Napoleon Bonaparte*. New York: Basic Books.

Baker, R. (2006), *Sperm Wars: Infidelity, sexual conflict, and other bedroom battles*. New York: Basic Books.

Baker, R. and Bellis, M. A. (1995), *Human Sperm Competition: Copulation, masturbation and infidelity*. London: Chapman & Hall.

Bandow, D. (1994), *The Politics of Envy: Statism as theology*. New Jersey: Transaction Publishers.

Barash, D. P. (2003), *The Survival Game: How game theory explains the biology of cooperation and competition*. New York: Owl Books.

Barash, D. P. and Barash, N. R. (2005), *Madame Bovary's Ovaries: A Darwinian look at literature*. New York: Delacourt Press.

Barkow, J. H., Cosmides, L. and Tooby, J. (1995), *The Adapted Mind: Evolutionary psychology and the generation of culture*. Oxford: Oxford University Press.

Barret, L. and Dunbar, R. (2009), *The Oxford Handbook of Evolutionary Psychology*. Oxford: Oxford University Press.

Barrett, D., Greenwood, J. G. and McCullagh, J. F. (2006), 'Kissing Laterality and Handedness', *Laterality*, 11 (6), pp. 573–9.

Baum, L. F. (2006), *The Wonderful Wizard of Oz*. New York: Signet Classics.

Belsky, S. (2010), *Making Ideas Happen: Overcoming the obstacles between vision and reality*. New York: Portfolio.

Benedict, R. (2006), *The Chrysanthemum and the Sword: Patterns of Japanese culture*. New York: Mariner Books.

Bjorklund, D. F. (2007) *Why Youth Is Not Wasted on the Young: Immaturity in human development* (Hoboken: Wiley-Blackwell).

Bloom, P. (2010), *How Pleasure Works: The new science of why we like what we like*. New York: W.W. Norton & Company.

Bloom, P. and Pinker, S. (1990), 'Natural Language and Natural Selection', *Behavioral and Brain Sciences*, 13, pp. 707–84.

Boyd, J. and Zimbardo, P. (2009), *The Time Paradox: The new psychology of time*. Rider.

Boynton, A., Fischer, B. and Bole, W. (2011). *The Idea Hunter: How to find the best ideas and make them happen*. San Francisco: Jossey-Bass.

Brizendine, L. (2007), *The Female Brain*. New York: Three Rivers Press.

Brizendine, L. (2010), *The Male Brain*. New York: Crown.

Brockman, J. (1995), *The Third Culture*. New York: Simon & Schuster.

Brockman, J. (2011), *Culture: Leading scientists explore civilizations, art, network, reputation, and the on-line revolution*. New York: Harper-Collins Publishers.

Brockman, J. (2013). *This Explains Everything: Deep, beautiful, and elegant theories of how the world works*. New York: Harper Perennial.

Brooks, D. (2001), *Bobos in Paradise: The new upper class and how they got there*. New York: Simon & Schuster.

Brotto, L. A., Knudson, G., Inskip, J., Rhodes, K. and Erskine, Y. (2010), 'Asexuality: A mixed-methods approach', *Archives of Sexual Behavior*, 39, pp. 599–712.

Brown, D. (1991), *Human Universals* (New York: McGraw-Hill).

Brown, S. (2009), *Play: How it shapes the brain, opens the imagination, and invigorates the soul*. New York: Penguin.

Burnham, T. and Phelan, J. (2000), *Mean Genes: From sex to money to food, taming our primal instincts*. London: Simon & Schuster.

Buss, D. M. (2003), *The Evolution of Desire: Strategies of human mating*. New York: Basic Books.

Buss, D. M. (2005), *The Handbook of Evolutionary Psychology*. Hoboken: John Wiley.

Buss, D. M. (2007), *Evolutionary Psychology: The new science of the mind*. Boston: Allyn and Bacon.

Buzan, B. and Hansen, L. (2009), *The Evolution of International Security Studies*. New York: Cambridge University Press.

Cain, S. (2013), *Quiet: The power of introverts in a world that can't stop talking*. London: Penguin Books.

Camus, A. (1991), *The Myth of Sisyphus, and Other Essays*. New York: Vintage.

Carrin, G. et al. (2001), *The Impact of the Degree of Risk-sharing in Health Financing on Health System Attaiment* (PDF), background paper, Working Group 3, WHO Commission on Macroeconomics and Health.

Cholle, F. P. (2012), *The Intuitive Compass: Why the best decisions balance reason and instinct*. San Francisco: Jossey-Bass.

Cohen, R. (2003), *The Good, the Bad and the Difference: How to tell right from wrong in everyday situations*. New York: Broadway Books.

Confer J. C., Easton, J. A., Fleischman, S., Goetz, C. D. and Buss D. M. (2010), 'Evolutionary Psychology: Controversies, questions, prospects, and limitations' *American Psychologist*, 65 (2), pp. 110–26.

Cooter, R. and Eldin, A. (2011), 'Law and Growth Economics: A framework for research' *Working Paper Series, Berkeley Program in Law and Economics*, UC Berkeley, 13 January 2011.

Cooter, R. D. and Schäfer, H.-B. (2012), *Solomon's Knot: How law can end the poverty of nations*. Princeton: Princeton University Press.

Coyne, J. A. (2010), *Why Evolution Is True*. Oxford: Oxford University Press.

Dawkins, R. (1976), *The Selfish Gene*. Oxford: Oxford University Press.

Dawkins, R. (2008), *The God Delusion*. New York: Mariner Books.

Dawkins, R. (2009), *The Greatest Show on Earth: The evidence for evolution*. New York: Free Press.

de Botton, A. (2002), *The Art of Travel*. New York: Vintage.

de Botton, A. (2004), *Status Anxiety*. New York: Pantheon Books.

de Botton, A. (2012), *How to Think More About Sex*. New York: The School of Life.

de Botton. A. (2013), *Religion for Atheists: A non-believer's guide to the uses of religion*. New York: Vintage.

de Grey, A. (2008), *Ending Aging: The rejuvenation breakthroughs that could reverse human aging in our lifetime*. New York: St Martin's Griffin.

De Mente, B. L. (1996), NTC*'s Dictionary of Mexican Cultural Code Words: The complete guide to key words that express how the Mexicans think, communicate, and behave*. New York: McGraw-Hill.

de Waal, F. (2005), *Our Inner Ape: A leading primatologist explains why we are who we are*. United States: Tantor Media.

de Waal, F. (2009), *Primates and Philosophers: How morality evolved*. Princeton: Princeton University Press.

Dennett, D. (1995), *Darwin's Dangerous Idea: Evolution and the meaning of life*. New York: Simon & Schuster Paperbacks.

Diamandis, P. H. and Kotler, S. (2012), *Abundance: The future is better than you think*. New York: Free Press.

Diamond, J. (1992), *The Third Chimpanzee: The evolution and future of the human animal*. New York: HarperCollins.

Diamond, J. (1997), *Guns, Germs, and Steel: The fates of human societies*. New York: W. W. Norton.

Diamond, J. (1997), *Why Is Sex Fun? The evolution of human sexuality*. New York: Perseus Books.

Diamond, J. (2003), *The Rise and Fall of the Third Chimpanzee: How our animal heritage affects the way we live*. New York: Vintage Books.

Drassinower, A. (2003), *Freud's Theory of Culture: Eros, loss, and politics*. Maryland: Rowman & Littlefield.

Dunn, K. M., Cherkas, L. F. and Spector, T. D. (2005), 'Genetic Influences on Variation in Female Orgasmic Function: A twin study', *Biology Letters*, 1 (3), pp. 260–63.

Eagleman, D. (2011), *Incognito: The secret lives of the brain*. New York: Pantheon Books.

Easterbrook, G. (2004), *The Progress Paradox: How life gets better while people feel worse*. New York: Random House.

Emden, C. J. (2005), *Nietzsche on Language, Consciousness, and the Body*. Champaign: University of Illinois Press.

Englund, S. (2004), *Napoleon: A political life*. Cambridge, MA: Harvard University Press.

Enriquez, J. (2001), *As the Future Catches You: How genomics and other forces are changing your life, work, health and wealth*. New York: Crown Business.

Estupinyà, P. (2013), *$S = ex^2$: La ciencia del sexo,* Barcelona: Random House Mondadori.

Etcoff, N. L. (2000), *Survival of the Prettiest: The science of beauty*. Michigan: Anchor.

Fang, F. C. and Casadevall, A. 'Why We Cheat', *Scientific American Mind*, May/June 2013.

Ferguson, N. (2012), *Civilization: The west and the rest*. New York: Penguin Books.

Fields, R. (1991), *The Code of the Warrior: In history, myth, and everyday life*. New York: Harper Collins.

Finkel, E. J., Eastwick, P. W., Karney, B. R., Reis, H. T. and Sprecher, S. (2012), 'Online dating: A critical analysis from the perspective of psychological science', *Psychological Science in the Public Interest*, February.

Fisher, H. E. (1992), *Anatomy of Love: The mysteries of mating, marriage and why we stray*. New York: Fawcett Columbine.

Fisher, H. E. (2000), *The First Sex: The natural talents of women and how they are changing the world*. New York: Ballantine Books.

Foer, J. (2012), *Moonwalking with Einstein: The art and science of remembering everything*. London: Penguin Books.

Forgas, J. P., Haselton, W. and Von Hippel, W. (2007), *Evolution and the Social Mind: Evolutionary psychology and social cognition*. Hove: Psychology Press.

Forster, E. M. (1964), *Abinger Harvest*. New York: Mariner Books.

Fortini, A. 'The Feminist Reawakening', *New York Magazine*, 13 April 2008.

Francken, A. B., Van de Wiel, H. B., Van Driel, M. F. and Weijmar Schultz, W. C. (2002), 'What Importance Do Women Attribute to the Size of the Penis?' *European Urology*, 42 (5), pp. 426–31.

Frank, R. (2007), *Richistan: A journey through the American wealth boom and the lives of the new rich*. New York: Three Rivers Press.

Frank, R. H. (1985), *Choosing the Right Pond: Human behavior and the quest for status*. New York: Oxford University Press.

Freud, S. (2000), *The Interpretation of Dreams*. Ware: Wordsworth Editions.

Friedman, T. L. (1999), *The Lexus and the Olive Tree: Understanding globalization*. New York: Farrar, Straus and Giroux.

Fromm, E. (1981), *On Disobedience, and Other Essays*. New York: Seabury Press.

Fromm, E. (1992), *The Anatomy of Destructiveness*. New York: Holt Paperbacks.

Gannon, M. J. and Pillai, R. (2010), *Understanding Global Cultures: Metaphorical journeys through 29 nations, clusters of nations, continents, and diversity*. Thousand Oaks: Sage Publications.

Garcia, J. R. and Reiber, C. (2008), 'Hook-up Behavior: A biopsychosocial perspective', *Journal of Social, Evolutionary, and Cultural Psychology*, 2, pp. 192–208.

Gilbert, D. (2007), *Stumbling on Happiness*. London: Harper Perennial.

Givens, D. B. (1978), 'The Nonverbal Basis of Attraction: Flirtation, courtship, and seduction', *Psychiatry*, 41 (4), pp. 346–59.

Gladwell, M. (2008), *Outliers: The story of success*. New York: Hachette.

Godin, S. (2003), *Purple Cow: Transform your business by being remarkable*. New York: Portfolio.

Godin, S. (2008), *Tribes: We need you to lead us*. New York: Portfolio.

Goleman, D. (1995), *Emotional Intelligence: Why it can matter more than IQ*. New York: Bantam Books.

Greenblatt, S., Meyer-Kalkus, R., Nyiri, P., Pannewick, F., Paul, H. and Zupanov, I. (2010), *Cultural Mobility: A manifesto*. Cambridge: Cambridge University Press.

Gueguen, N. (2007), 'Women's Bust Size and Men's Courtship Solicitation', *Body Image*, 4 (4), pp. 386–90.

Gueguen, N. (2008), 'The Effect of a Woman's Smile on a Man's Courtship Behavior', *Social Behavior and Personality*, 36, pp. 1,233–6.

Gueguen, N. (2011), 'The effect of women's suggestive clothing on men's behavior and judgment: A field study', *Psychological Reports*, 109 (2), p. 635–8.

Hall, E. T. (1959), *The Silent Language*. New York: Anchor Books.

Hall, E. T. (1966), *The Hidden Dimension*. New York: Anchor Books.

Hall, E. T. (1989), *Beyond Culture*. New York: Anchor Books.

Hall, E. T. (1990), *Hidden Differences: Doing business with the Japanese*. New York: Anchor Books.

Hall, E. T. (1990), *Understanding Cultural Differences: Germans, French and Americans*. Boston: Nicholas Brealey Publishing.

Hamann, S., Herman, R. A., Nolan C. L. and Wallen, K. (2004), 'Men and Women Differ in Amygdale Response to Visual Sexual Stimuli', *Nature Neuroscience,* 7 (4), pp. 411–16.

Hamer, D. H. and Copeland, P. (1999), *Living With Our Genes: Why they matter more than you think*. New York: Anchor Books.

Harris, S. (2012), *Free Will*. New York: Free Press.

Harrison, L. E. and Huntington, S. P. (2006), *Culture Matters: How values shape human progress*. New York: Basic Books.

Hauser, M. D. (2006), *Moral Minds: The nature of right and wrong*. New York: Harper Perennial.

Heath, C. and Heath, D. (2013), *Decisive: How to make better choices in life and work*. New York: Crown Business.

Henley, N. M. and Goodchilds, J. D. (2000), *The Curmudgeon Woman*. Missouri: Andrews McMeel Publishing.

Hermann, H. (2000), *Demian*. New York: Dover Publications.

Hill, S. E. and Buss, D. M. (2008), 'The Mere Presence of Opposite-Sex Others on Judgments of Sexual and Romantic Desirability: Opposite effects for men and women', *Personality and Social Psychology Bulletin,* 34 (5), pp. 635–47.

Hilton, D. L., Jr. and Watts, C. (2011), 'Pornography Addictions: A neuroscience perspective', *Surgical Neurology International,* 2, p. 19.

Hofstede G., Hofstede, G. J. and Minkov, M. (2010), *Cultures and Organizations: Software of the mind, international cooperation and its importance for survival*. New York: McGraw-Hill.

Hofstede G., Hofstede, G. J. and Pedersen, P. B. (2002), *Exploring*

Culture: Exercises, stories and synthetic cultures. Boston: Intercultural Press.

Honoré, C. (2004), *In Praise of Slow: How a worldwide movement is challenging the cult of speed*. London: Orion.

Horner, J. and Gorman, J. (2010), *How to Build a Dinosaur: The new science of reverse evolution*. New York: Plume Books.

Huntington, S. (1977), *The Clash of Civilizations and the Remaking of the World Order*. New York: Simon & Schuster.

Jackendoff, R. and Pinker, S. (2005), 'The Faculty of Language: What's special about it?' *Cognition*, 95, pp. 201–36.

Jacobi, J. (1942), *Psychology of C. G. Jung*. London: Routledge & Kegan Paul.

James, W. (1950), *The Principles of Psychology*. New York: Dover Publications.

Janet, L. Peplau, L. A. and Frederick, D. A. (2006), 'Does Size Matter? Men's and women's views on penis size across the lifespan', *Psychology of Men and Masculinity*, 7 (3), pp. 129–43.

Jensen, A. R. and Miele, F. (2002), *Intelligence, Race, and Genetics: Conversations with Arthur R. Jensen*. Boulder: Westview Press.

Johnson, S. (2005), *Mind Wide Open: Your brain and the neuroscience of everyday life*. New York: Scribner.

Johnson, S. (2012), *Future Perfect: The case for progress in a networked age*. New York: Riverhead Books

Jung, C. (1968), *The Archetype and the Collective Unconscious*. Princeton: Princeton University Press.

Jung, C. (1968), *Man and His Symbols*. New York: Dell.

Jung, C. and Storr, A. (ed.) (1999), *The Essential Jung: Selected writings*. Princeton: Princeton University Press.

Kahneman, D. (2011), *Thinking, Fast and Slow*. New York: Farrar, Straus and Giroux.

Kaku, M. (2012), *Physics of the Future: The inventions that will transform our lives*. London: Penguin Books.

Kanter, B. and Fine, A. H. (2010), *The Networked Nonprofit: Connecting with social media to drive change*. San Francisco: Jossey-Bass.

Kay, J. (2004), *Culture and Prosperity: Why some nations are rich but most remain poor*. New York: HarperBusiness.

Kelling, G. L. and Wilson, J. Q (1982). 'Broken Windows', *The Atlantic*, March.

Kirshenbaum, S. (2011), *The Science of Kissing*. New York: Hachette.

Knight, J. (1968), *For The Love of Money: Human behavior and money*. Philadelphia: J. B. Lippincott.

Kristof, N. D. and WuDunn, S. (2010), *Half the Sky: How to change the world*. London: Virago.

Krugman, P. (2013), *End This Depression Now!* New York: W. W. Norton.

Kruzban, R. (2012), *Why Everyone (Else) Is a Hypocrite: Evolution and the modular mind*. Princeton: Princeton University Press.

Kurzweil, R. (2005), *The Singularity Is Near: When humans transcend biology*. New York: Viking.

Landes, D. (1999), *The Wealth and Poverty of Nations: Why some are so rich and some so poor*. New York: W. W. Norton.

Lee, L., Loewenstein, G., Ariely, D., Hong, J. and Young, J. (2008), 'If I'm Not Hot, Are You Hot or Not? Physical-attractiveness evaluations and dating preferences as a function of one's own attractiveness', *Psychological Science*, 19 (7), pp. 669–77.

Lee Kuan Yew (2000), *From Third World to First: The Singapore story, 1965–2000*. New York: HarperCollins.

Lehrer, J. (2009), *The Decisive Moment: How the brain makes up its mind*. Edinburgh: Canongate Books.

Levitt, S. (2006), *Freakonomics: A rogue economist explores the hidden side of everything*. London: Penguin Books.

Leyner, M. and Goldberg, B. (2006), *Why Do Men Fall Asleep After Sex?: More questions you'd only ask a doctor after your third whiskey sour*. New York: Random House.

Linden, D. J. (2012), *The Compass of Pleasure: How our brains make fatty foods, orgasm, exercise, marijuana, generosity, vodka, learning, and gambling feel so good*. New York: Penguin.

Livermore, D. (2009), *Cultural Intelligence: Improving your CQ to engage our multicultural world*. Grand Rapids: Baker Academic.

Livermore, D. (2009), *Leading with Cultural Intelligence: The new secret to success*. New York: AMACON.

MacHale, D. (2003), *Wit*. Kansas City: Andrews McMeel.

MacLean, Paul D. (1990), *The Triune Brain in Evolution: Role in paleocerebral functions*. New York: Plenum Press.

MacLeod, H. (2009), *Ignore Everybody: And 39 other keys to creativity*. New York: Portfolio.

Mahbubani, K. (2013), *The Great Convergence: Asia, the west, and the logic of one world*. New York: PublicAffairs.

Manning, J. (2007), *The Finger Book*. London: Faber and Faber.

Marcus, G. (2009), *Kluge: The haphazard evolution of the human mind*. London: Faber and Faber.

Markman, A. B. (2012), *Smart Thinking: Three essential keys to solve problems, innovate, and get things done*. New York: Perigee.

Marmot, M. (2004), *The Status Syndrome: How social standing affects our health and longevity*. New York: Owl Books.

Maslow, A. H. (1943), 'A Theory of Human Motivation', *Psychological Review*, 50, pp. 370–96.

May, E. R. (1986), *Knowing One's Enemies: Intelligence assessment before the two world wars*. Princeton: Princeton University Press.

May, R. (2008), *Love and Will*. Princeton: Recording for the Blind and Dyslexic.

Medina, J. (2009), *Brain Rules: 12 principles for surviving and thriving at work, home and school*. New York: Pear Press.

Meston, C. M. (2000), 'Sympathetic Nervous System Activity and Female Sexual Arousal', *American Journal of Cardiology*, 86 (2/1), pp. 30–34.

Meston, C. M. and Buss, D. M. (2009), *Why Women Have Sex:*

Understanding sexual motivations, from adventure to revenge (and everything in between). New York: Times Books.

Miller, G. (2009), *Spent: Sex, evolution, and consumer behavior*. New York: Viking.

Milner, M. (1994), *Status and Sacredness: A general theory of status relations and an analysis of Indian culture*. New York: Oxford University Press.

Mischel, W., Shoda, Y. and Rodriguez, M. L. (1989), 'Delay of Gratification in Children', *Science*, 244, pp. 933–8.

Moore, M. M. (2010), 'Human Nonverbal Courtship Behavior: A brief historical review', *Journal of Sex Research*, 47 (2–3), pp. 171–80.

Morris, D. (1994), *The Human Animal: A personal view of the human species*. London: BBC Books.

Morris, D. (1996), *The Human Zoo: A zoologist's classic study of the urban animal*. New York: Kodansha International.

Moyo, D. (2009), *Dead Aid: Why aid is not working and how there is a better way for Africa*. New York: Farrar, Straus and Giroux

Moyo, D. (2013), *Winner Take All: China's race for resources and what it means for us*. London: Penguin Books.

Myers, I. B. and Myers P. B. (1995), *Gifts Differing: Understanding personality type*. Boston: Nicholas Brealey.

Nisbett, A. (1976), *Konrad Lorenz*. London: Dent.

Nolan, P. (2004), *Transforming China: Globalization, transition and development*. New York: Anthem Press.

Nussbaum, M. C. (2011), *Creating Capabilities: The human development approach*. Cambridge, MA: Harvard University Press.

Parada, M., Vargas, E. B., Kyres, M., Burnside, K. and Pfaus, J. G. (2012), 'The Role of Ovarian Hormones in Sexual Rewards States of the Female Rat', *Hormones and Behavior*, 62 (4), pp. 442–7.

Paul, E. L. and Hayes, K. A. (2002), 'The Casualties of "Casual" Sex: A qualitative exploration of the phenomenology of college students' hook-ups', *Journal of Social and Personal Relationships*, 19, pp. 639–61.

Paul, E. L., McManus, B. and Hayes, A. (2000), '"Hook-ups": Characteristics and correlates of college students' spontaneous and anonymous sexual experience', *Journal of Sex Research*, 37, pp. 76–88.

Payer, L. (1996), *Medicine and Culture: Varieties of treatment in the United States, England, West Germany, and France*. New York: Henry Holt and Company.

Petersen, J. L. and Hyde, J. S. (2010), 'A Meta-Analytic Review of Research on Gender Differences in Sexuality, 1993–2007', *Psychological Bulletin*, 136 (1), pp. 149–65.

Petrie, D. and Kam-Ki Tang (2008), 'A Rethink on Measuring Health inequalities Using the Gini Coefficient', School of Economics Discussion Paper No. 381, University of Queensland.

Pinker, S. (1994), *The Language Instinct*. New York: Morrow.

Pinker, S. (2003) *The Blank Slate: The modern denial of human nature*. New York: Penguin Books.

Pinker, S. (2009), *How the Mind Works*. New York: W. W. Norton.

Pinker, S. (2011), *The Better Angels of Our Nature: Why violence has declined*. New York: Viking.

Pool, R. (1994), *Eve's Rib: Searching for the biological roots of sex differences*. New York: Crown Publishers.

Posner, R. A. (1992), *Sex and Reason*. Cambridge, MA: Harvard University Press.

Provine, R. R. (2012), *Curious Behavior: Yawning, laughing, hiccupping, and beyond*. Cambridge, MA: Belknap Press.

Punset, E. (2007), *El alma está en el cerebro: Radiografía de la máquina de pensar*. Mexico: Punto de Lectura.

Punset, E. (2010), *El viaje al poder de la mente*. Barcelona: Destino.

Punset, E. (2010), *Por qué somos como somos*. Mexico: Punto de Lectura.

Punset, E. (2013), *Viaje al optimismo: Las claves del futuro*. Mexico City: Editorial Diana.

Putnam, R. D. (2000), *Bowling Alone: The collapse and revival of American community*. New York: Simon and Schuster.

Randall, L. (2012), *Knocking on Heaven's Door: How physics and scientific thinking illuminate our universe*. London: Vintage Publishing.

Rapaille, C. (1972), *Laing*. Paris: Editions Universitaires.

Rapaille, C. (1973), *La Relation créatrice*. Paris: Éditions Universitaires.

Rapaille, C. (1975), 'Wisdom Of Madness', thesis, Michigan State University.

Rapaille, C. (1976), *La Communication créatrice*. Paris: Éditions Dialogues.

Rapaille, C. (1978), *Si vous écoutiez vos enfants*. Paris: Mengès.

Rapaille, C. (1980), *Le Trouple*. Paris: Mengès.

Rapaille, C. (1981), *Escúchelo: Es su hijo*. Barcelona: Pomaire.

Rapaille, C. (1982), *Comprendre ses parents et ses grands parents*. Paris: Marabout.

Rapaille, C. (1984), *Versteh' deine Eltern*. Munich: Bucher Verlag.

Rapaille, C. (2001), *Seven Secrets of Marketing in a Multi-Cultural World*. Utah: Executive Excellence.

Rapaille, C. (2003), *Social Cancer: Decoding the archetype of terrorism*. New York: Tuxedo Productions.

Rapaille, C. (2006), *The Culture Code: An ingenious way to understand why people around the world live and buy as they do*. New York: Crown Business.

Reich, W. (1973), *The Function of the Orgasm*. New York: Farrar, Straus and Giroux.

Restak, R. (2006), *The Naked Brain: How the emerging neurosociety is changing how we live, work, and love*. New York: Harmony Books.

Revonsuo, A. (2006), *Inner Presence: Consciousness as a biological phenomenon*. Cambridge, MA: MIT Press.

Riding, A. (1987), *Vecinos distantes: Un retrato de los mexicanos*. Mexico: Planeta.

Ridley, M. (1993), *The Red Queen: Sex and the evolution of human nature*. London: Viking.

Ridley, M. (1996), *Evolution*. Cambridge, MA: Blackwell Science.

Ridley, M. (2011), *The Rational Optimist: How prosperity evolves*. New York: Harper Perennial.

Rilling, J. K. Scholz, J., Preuss, T. M. Glasser, M. F., Errangi, B. K. and Behrens, T. E. (2012), 'Differences between Chimpanzees and Bonobos In Neural System Supporting Social Cognition', *Social and Cognitive Affective Neuroscience*, 7 (4), pp. 369–79.

Robinson, K. (2013), *Finding Your Element: How to discover your talents and passions and transform your life*. London: Allen Lane.

Roemer, A. (1995), *El juego de la negociación*. Mexico: ITAM.

Roemer, A. (1997), *Economía y derecho: Políticas públicas*. Mexico: Porrúa.

Roemer, A. (1998), *Sexualidad, derecho y políticas públicas*. Mexico: Porrúa.

Roemer, A. (2000), *Derecho y economía: Una revision de la literatura*. Mexico: FCE-CED.

Roemer, A. (2001), *Economía del Crimen*. Mexico: Limusa.

Roemer, A. (2003), *Enigmas y paradigmas: Una exploración entre el arte y la política pública*. Mexico: Limusa.

Roemer, A. (2005), *Entre lo publico y lo privado: 1300 + 13 preguntas para pensar sobre pensar*. Mexico: Editorial Noriega.

Roemer, A. (2005), *Felicidad: Un enfoque de derecho y economía*. Mexico: Instituto de Investigaciones Jurídicas UNAM.

Roemer, A. (2006), *Terrorismo y crimen organizado: Un enfoque de derecho y economía*. Mexico: Instituto de Investigaciones Jurídicas UNAM.

Roemer, A. (2007), *No: Un imperativo de la generacion next*. Mexico: Santillana Ediciones Generales.

Roemer, A. (2008), *El otro Einstein*. Mexico: Porrúa.

Roemer, A. (2011), *Oskar y Jack*. Mexico: Porrúa.

Roemer, A. and Ghersi, E. (2008), *¿Por qué amamos el futbol?: Un enfoque de derecho y economía*. Mexico: Porrúa.

Roheim, G. (1943), *The Origin and Function of Culture*. New York: Nervous and Mental Disease Monographs.

Rosenfield, I. (1988), *The Invention of Memory*. New York: Basic Books.

Russell, B. (1983), *The Will to Doubt*. New York: Philosophical Library.

Russell, B. and Ruse, M. (1997), *Religion and Science*. New York: Oxford University Press.

Ryan, C. and Jetha, C. (2010), *Sex at Dawn: How we mate, why we stray, and what it means for modern relationships*. New York: HarperCollins.

Sachs, J. (2005), *The End of Poverty: Economic possibilities for our time*. New York: Penguin Press.

Sahrot, T. (2011), *The Optimism Bias: A tour of the irrationality positive brain*. New York: Pantheon.

Sapolsky, R. M. (1998), *The Trouble with Testosterone: And other essays on the biology of the human predicament*. New York: Touchstone.

Sapolsky, R. M. (2002), *A Primate's Memoir: A neuroscientist's unconventional life among the baboons*. New York: Simon & Schuster.

Sapolsky, R. M. (2004), *Why Zebras Don't Get Ulcers*. New York: Owl Books.

Sapolsky, R. M. (2005), *Monkeyluv: And other essays on our lives as animals*. New York: Scribner.

Schacter, D. L. (2002), *The Seven Sins of Memory: How the mind forgets and remembers*. New York: Mariner Books.

Schein, E. H. (2009), *The Corporate Culture Survival Guide*. San Francisco: John Wiley.

Schellnhuber, H. J. et al. (2010), *Global Sustainability: A Nobel cause*. Cambridge: Cambridge University Press.

Schmitt, D. P. and Pilcher, J. J. 'Evaluating Evidence of Psychological Adaptation: How do we know one when we see one?', *Psychological Science*, 15, p. 643.

Schopenhauer, A. (1995), *Essays on Freedom of the Will*. New Jersey: Wiley-Blackwell.

Schwartz, B. (2004), *The Paradox of Choice: Why more is less*. New York: Harper Perennial.

Segev, I. and Markram, H. (2011), *Augmenting Cognition*. Lausanne: EPFL Press.

Senor, D. and Singer, S. (2011), *Start-up Nation: The story of Israel's economic miracle*. New York: Twelve.

Shah, J. and Christoper, N. (2002), 'Can Shoe Size Predict Penile Length?', BJU *International*, 90 (6), pp. 586–7.

Shermer, M. (2011), *The Believing Brain: From ghosts and gods to politics and conspiracies – how we construct beliefs and reinforce them as truths*. New York: Times Books.

Smith, D. L. (2004), *Why We Lie: The evolutionary roots of deception and the unconscious mind*. New York: St Martin's Press.

Sober, E. and Wilson, D. S. (1999), *Unto Others: The evolution and psychology of unselfish behavior*. Cambridge, MA: Harvard University Press.

Spencer, H. (1864). *Principles of Biology*. (London: Williams and Norgate)

Spitz, R. A. (1945), 'Hospitalism: An inquiry into the genesis of psychiatric conditions in early childhood', *Psychoanalytic Study of the Child*, 1, pp. 53–74.

Spurlock, M. (2005), *Don't Eat This Book*. New York: Penguin.

Standage, T. (2006), *A History of the World in 6 Glasses*. New York: Walker.

Stevens, A. (1983), *Archetypes: A pioneering investigation into the biological basis of Jung's theory of archetypes*. New York: Quill.

Sukel, K. (2012), *This Is Your Brain on Sex: The science behind the search for love*. New York: Simon & Schuster Paperbacks.

Sunstein, C. R. and Thaler, R. H. (2008), *Nudge: Improving decisions about health, wealth, and happiness*. New Haven: Yale University Press.

Swami, V. (2010), 'The Attractive Female Body Weight and Female Body Dissatisfaction in 26 Countries Across 10 World Regions: Results of the International Body Project I', *Personality and Social Psychology Bulletin*, 36, p. 309.

Swami, V. and Tovée, M. J. (2006), 'Does Hunger Influence Judgments of Female Physical Attractiveness?', *British Journal of Psychology*, 97 (3), pp. 353–63.

Taleb, N. N. (2012), *Antifragile: Things that gain from disorder*. London: Penguin Books.

Trivers, R. (2011), *The Folly of Fools: The logic of deceit and self-deception in human life*. New York: Basic Books.

Veblen, T. (2005), *The Theory of the Leisure Class: An economic study of institutions*. Delhi: Aakar Books.

Vicente, P. C. and Kaufmann, D. (2011), 'Legal Corruption', *Economics and Politics*, 23(2), pp. 195–219.

Vogel, E. F. (2011), *Deng Xiaoping and the Transformation of China*. Cambridge, MA: Harvard University Press.

Wallen, K. and Lloyd, E. A. (2011), 'Female Sexual Arousal: Genital anatomy and orgasm in intercourse', *Hormones and Behavior*, 59 (5), pp.780–92.

Warren, F. (2007), *A Lifetime of Secrets: A PostSecret book*. New York: William Morrow.

Weber, M. (2001), *The Protestant Ethic and the Spirit of Capitalism*. London: Routledge.

Wilson, E. O. (2000), *Sociobiology: The new synthesis*. Cambridge, MA: Belknap Press.

Wilson, J. Q. and Herrnstein, R. J. (1985), *Crime and Human Nature: The definitive study of the causes of crime*. New York: Free Press.

Winston, R. (2002), *Human Instinct: How our primeval impulses shape our modern lives*. London: Bantam Books.

Wobber, V., Hare, B., Maboto, J., Lipson, S., Wrangham, R. and Ellison, P. T. (2010), 'Differential Changes in Steroid Hormones Prior to Competition in Bonobos and Chimpanzees', *Proceeding of the National Academy of Science,* 107 (28), pp. 457–62.

Wright, P. J. (2013), 'U.S. Males and Pornography, 1973–2010: Consumption predictors, correlates', *Journal of Sex Research,* 50 (1), pp. 60–71.

Wright, R. (1995), *The Moral Animal: Why we are the way we are – the new science of evolutionary psychology.* New York: Vintage.

Zamoyski, A. (2005), *Moscow 1812: Napoleon's fatal march.* New York: Harper Perennial.

Zuckerman, M. (1992), *Incredibly American.* Milwaukee: ASQC Quality Press.

Glossary

archetype A pre-existing structure or pattern that enables human beings to meet their biological needs and understand their human condition. It is imprinted at an early age when, as children, we discover and experience the external world. There are two kinds of archetype: the Jungian universal archetype, which we call 'schema' (plural: 'schemata'), and the cultural archetype. In this book, the term 'archetype' refers to the latter.

archetypes, universal Universal archetypes (e.g. 'mother') are related to the biological structures or schemata that pre-organize human life. They are imprinted in humans wherever that species exists. These mental and behavioural forms are present in all members of the species but develop uniquely in each individual. Together, the universal archetypes make up the collective unconscious (q.v.).

archetypes, cultural Cultural archetypes are those structures that are shared by a single culture. They are the cultural grammar that gives meaning and significance to any element of a given culture, at a given time. Every element of a culture (a word, object, role, concept, etc.) has an archetype. Cultural archetypes are inherited but are not genetically transmitted. They pre-organize our perceptions of the world and the way we react to it. They differ from one culture to another but are common to all the members of a culture or subculture. Archetypes are empty – they are waiting to be expressed. They are composed of dualistic forces that are always in action, always dynamic. An archetype is the force field that organizes everything that happens in a given culture.

code The code is a simple, synthetic way of understanding the order and meaning of the archetype (e.g., 'the cheese is dead' is the American code for cheese). The code is the access to the structure of the forces that organize a culture (the way we think and function); it is a system of signals by which a convention is transmitted. The code is like the combination for a safe's lock. If you have all the numbers, but not the proper order, you cannot open the door. The cultural codes that are the subject of this book are unconscious. Bringing them into awareness – decoding them – can illuminate the psyche and help explain why we are the way we are.

crystallization The process by which a cultural archetype assumes permanence and definition. Crystallization of archetypes enables a culture to be transmitted from one generation to the next. Language is the first step in crystallization. The imprint receives a name (label). Whenever that label is used, the imprint becomes deeper, even when the original experience has faded from consciousness. Cultural archetypes gradually become crystallized into norms and rules, and finally into laws.

cultural shadow Just as C. G. Jung spoke of the shadow as the hidden side of a personality (Dr Jekyll and Mr Hyde), the cultural shadow is the dark or hidden side of a culture. We have to understand this dark side and help the members of the culture to become aware of the shadow and control it. For example, in America the John Wayne figure represents the archetype of Americans as morally good, strong and incorruptible. But the cultural shadow is always present in such figures as Jimmy Swaggart or Charles Manson.

culture The whole body of socially transmitted behaviours, beliefs, feelings, etc., which ensures that a specific population will survive and endure. A culture provides the pre-existing structures or common patterns (archetypes) that enable a group of people to meet their biological needs and understand the human condition. The structures that evolve and are transmitted in one culture will be distinct from those in any other.

forces Forces – drives, attractions, repulsions – are the way a culture deals with its biological needs and ensures its own survival. Forces are culturally specific, even if some forces are the same in several cultures. Forces are the strong magnetic attractions that create an archetype's dualistic character. Every cultural archetype is composed of two forces, pulling in opposite directions. (For example, if there is a force pulling towards diversity, it will be opposed by a force pulling towards uniformity.) It is this energy to attract and repel that causes movement (life) in a culture.

imprint The permanent impression made on our unconscious by the initial experience of an archetype and its accompanying emotion. This learning process cannot occur in the absence of emotion, which provides the energy for imprinting (emotion causes the brain to release neurotransmitters, which establish the learning connection – the neutral highway). From that point – the imprinting moment – the imprint will be repeatedly reinforced, and will strongly condition our thought processes, emotions, etc. There are both biological and cultural imprints; the latter vary from culture to culture and reside in the cultural collective unconscious.

imprinting The rapid formation of an attachment between a human being and an environmental 'object'. This object could be a thing, a person or a concept – any element of the world. Konrad Lorenz originally described imprinting in 1952. He found, for example, that a newly hatched duckling would form an attachment to any kind of moving object and follow it as its mother, provided that the object was present at just the right moment in the duckling's life. Lorenz demonstrated attachments between ducklings and a wooden box on wheels, a human being and a bird of a different species. Lorenz also found that imprinting occurs only during a critical period, which we call a window in time, after which it is difficult, if not impossible, to imprint. Emotion is the energy necessary to imprint the neural pathways (mental highways). The sooner an imprint occurs, the stronger it is and the longer it will last. An imprint related to survival and engraved at a very early age will last forever if regularly maintained and reinforced.

instinct The inborn capacity to perform a complex behavioural task. Sometimes called species-specific behaviour, in contrast to universal archetypes, which are cultural-specific organizations of behaviour that are acquired (learned) and transmitted by parents or substitutes.

latent structure The other side of the coin. The latent structure is the unconscious dimension of the archetype's code. Although unexpressed or unknown, it exists as part of the whole, and, once we become aware of it, it completes our understanding of reality. The latent structure is what the archetype process seeks to discover.

mental highways The mental connections that have been imprinted, reinforced and maintained, and which represent the collective software that people use to relate to and understand the world. When we are born we have billions of potential connections in the neural network of our brain, but we have no telephone numbers. Learning through imprints creates the telephone lines or numbers that are available in a given culture and are the most frequently used by the members of the culture. The cars that use the highways are called clichés, stereotypes, icons, heroes, etc.

ritual This refers to an unconscious code that is embodied in a system of behaviour. This codified system condenses the hero's journey into a form that ordinary humans can repeat. Rituals are the way people establish a relationship with spiritual or divine powers or with the secular substitutes for those powers (e.g. ideals). The ritual is repeated, and retains its sacred dimensions, even when it no longer has links to its origins. How members of a culture greet one another, which hand they use if they shake hands, etc., is based on some long-forgotten ritual. Unlike a custom, a ritual refers directly to latent forces and to a certain sacredness or inviolability. Everyday, mundane routines can be described as rituals, and these take on added importance in cultures where religious rituals are less prominent.

schema (plural, **schemata**) A universal archetype (Jung), a basic biological structure for survival. A schema is the potential to act in certain ways in order to survive. For example, the 'grasping schema' refers to the general ability to grasp things; it is the cognitive structure that makes all acts of grasping possible. A schema can also be thought of as an element in an organism's cognitive structure.

script The personal, individual life structure. Every script is unique. For the most part we are not aware of our individual scripts, which is why we keep repeating them. For example, a person bounces from one job to another, always blaming their boss, the company, the environment, and never examining their own script for the real roots of their dissatisfaction. The purpose of psychoanalysis is to make people aware of their script, to free them from repeating it indefinitely.

stereotype A person or thing that is thought to typify a well-known and predictable pattern of behaviour, as in, 'professors are absent-minded'. A stereotype is not factual, but it nonetheless rings true with people because they recognize a grain of truth in it.

structure The organization of elements, which is independent of them. For example, melody is independent of the notes that make it up; a triangle is independent of the three elements that shape it (they could be three fruits, three stones, three pencils). An information structure is what is transmitted when two human beings, one male and one female, have a child. The child is predictably a human being (instead of a bird or a fish) because the parents have transmitted the information structure, i.e. human being. The information content (blue eyes, brown hair, etc.) varies, which is what makes every human being unique.

symbol An expression or manifestation of the logic of emotion, which it synthesizes and concentrates. It is charged with emotion and meaning. Symbols are imprinted at an early age, and, from that point on, the symbolic meaning is inseparable from the word or object or concept it represents. Symbols connect all the people who share the same imprints. They are sometimes the link between the conscious and the unconscious worlds. Among Jews, the Star of David is a symbol of Judaism; among motorists, a stop sign symbolizes the authority of a traffic cop. Ritual practices are symbolic because they make reference to an absent relationship (e.g. Christian communion, with the wine and bread representing the blood and body of Christ).

tension Tension describes the relationship between elements. In archetypes it is the condition created by the bipolarity of forces – two forces of an axis pulling in opposite directions. A pair of axes (a quaternity) represents a magnetic field of tensions. Forces that are far apart on an axis create extreme tension; those that are closer together create more modest tension. Tensions are what make a culture dynamic. They create the movement that gives a culture life.

unconscious That part of an individual's or culture's psyche that cannot be directly observed. The unconscious is composed of those forces that influence our personality and behaviour that are out of awareness. It encompasses everything that is not conscious.

unconscious, individual The concept of the individual unconscious originated in the eighteenth century but was explored and popularized a century later by Sigmund Freud, who called it the libido or life force. Psychoanalysis is the methodology for bringing the individual unconscious into awareness.

unconscious, collective The collective unconscious is a term that was first used by Carl G. Jung to explain all forms of biological survival. He believed it was composed of universal archetypes, pre-existing structures or patterns (gestalts) that are present in all members of the species and that enable that species to meet its biological needs for survival. To avoid confusion, we refer to these universal archetypes as schemata.

unconscious, collective cultural This is the link between the individual and the collective unconscious, and it is the subject of our exploration. The collective cultural unconscious is where all the shared imprints of a culture – the cultural archetypes – reside. The cultural collective unconscious is what all members of a culture have available to them to function and survive in their culture. The archetypal structures can explain behaviour once they are brought into awareness. The discovery of archetypes is in effect the psychoanalysis of a culture's unconscious.

APPENDICES

Appendix 1: Building the Index

How Is the R^2 Index Compiled?

1.1 Survival

For the Survival component we used the Where To Be Born Index, developed by *The Economist* Intelligence Unit and published by *The Economist*, as well as overall expenditure on health per capita and expenditure on education as a percentage of Gross National Income.

$$\textbf{Survive} = \frac{\text{Where to be born Index} + (\text{Expenditure on health} + \text{Expenditure on education})}{2}$$

The Where To Be Born Index was published in 2013. It covers eighty countries, and considers eleven indicators. Some are fixed, such as geographic location, whereas others change slowly over time (demography, social and cultural characteristics). It considers the Quality-of-Life Index, which links the results of subjective life-satisfaction surveys – how happy people say they are – to objective determinants of the quality of life across countries. This index considers GDP per capita as well as crime, trust in public institutions and the quality of family life. For the R^2 Index, we standardize the Where To Be Born Index from 0 to 1.

Why Consider Expenditure in Health and Education?

It is important to understand the impact of public policy on development strategies, given that governments assign a significant proportion of their budget to this. *One would therefore expect that citizens should have more opportunities for educational and health development.*

In order to obtain a broader sample for the R^2 Index, we selected those variables with the greatest correlation to social mobility, using Pearson's coefficient. This measure reflects the degree of linear relationship between two variables; its range goes from −1 to 1. When a correlation has the value of −1, it means a perfect inverse linear relationship between variables.*

The value for social mobility was taken from an index constructed by Fundación Espinosa Rugarcía,† which has values from 0 to 1, where 1 means perfect social mobility and 0 nil social mobility. The variables selected for the correlations were expenditure in health and expenditure in education as a percentage of GDP.

A positive gradient is observed in this graph, implying that there is a direct relationship between share of GDP in education and health, and social mobility. It means that the higher the percentage of GDP devoted to social expenditure (education and health), the better the opportunities for social mobility within that country. In addition, Pearson's correlation has a high value, meaning that there is a strong relationship between both variables.

The results take into account only the economic and developmental variables for social mobility, without considering the

* D. Gujarati, *Basic Econometrics* (Boston: McGraw Hill, 4th edn, 2003).

† http://www.fundacionesru.org/

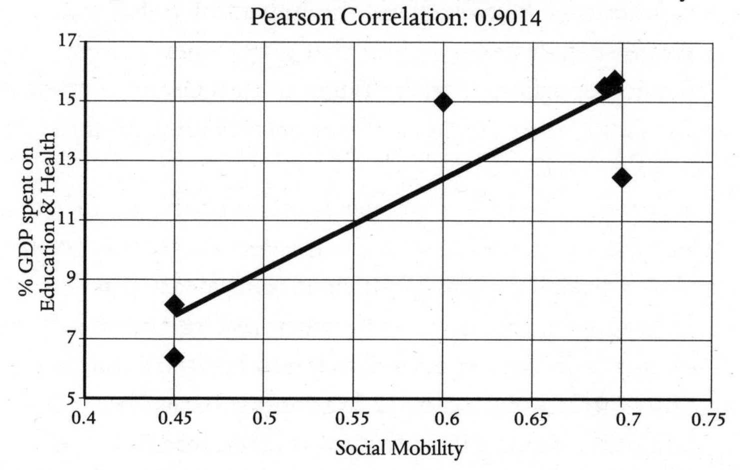

countries' culture codes. It is important to bear in mind that the values in this graph do not show the country's economic progress; instead, they show important economic and development variables, which are essential for social mobility.

In Survival, the country that has the greatest value is Denmark, followed closely by Switzerland, whereas the United States is in fourth place. Those with the lowest values are Bangladesh and Pakistan. Of the top ten countries with the highest values for Survival, seven are European, two North American (Canada, the US) and the other is Australia. At the bottom of our sample, Russia, India and the Philippines stand out as having low values (not the lowest ones, though) for Survival.

1.2 *Sex*

For the Sex component, we used the Gender Inequality Index (GII) trends, 1995–2011, recorded by the United Nations for 146 countries.

The GII reflects women's disadvantages in three dimensions:

- reproductive health
- empowerment
- participation in the labour force.

This index shows the loss in human development due to inequality between female and male achievements in these three dimensions.

- *The reproductive health dimension* is measured by two indicators: maternal mortality ratio and adolescent fertility rate. Maternal mortality is the result of poor or absent professional care during pregnancy, and of health problems related to fertility. This indicator shows us how mothers care for their own health during pregnancy, even when they do not receive professional help. Thus, we refer to the mother's own survival instinct and her instinctive protection for her children. In addition, family planning, women's rights and sexual information is captured in the adolescent fertility rate.
- *The empowerment dimension* is also measured by two indicators: the share of parliamentary seats held by each sex, and by both secondary and higher education attainment levels.
- *The labour force dimension* is measured by women's participation in the workforce. The more women participate, the more open a society is to equalizing rights regardless of gender.

The Gender Inequality Index is designed to reveal the extent to which national achievements in these aspects of human development are eroded by gender inequality, and to provide empirical foundations for policy analysis and advocacy efforts. It ranges from 0, which indicates that women and men fare equally, to 1, which indicates that women fare as poorly as possible in all measured dimensions.

For the purposes of our R^2 index, we used the results of the GII as a contraindication for moving up, so we used the formula: sex = 1 – (GII).

Viewed thus, the results range from 0 to 1, the same as all the components in our index, meaning that the closer we get to 1, the better for mobility it will be. In this case, the more the gender inequality gap closes, the better for moving up. The countries with the highest values have the smallest gap. We are talking about Sweden, the Netherlands, and again Denmark and Switzerland. The lowest value for Sex goes to Saudi Arabia, but Morocco, Bangladesh, Pakistan, India and Kenya are not far behind.

1.3 Security

$$\textbf{Security} = \frac{\text{Index of Economic Freedom} + \text{Human Security Index}}{2}$$

For the Security component we used two indexes: the *Index of Economic Freedom* and the Human Security Index. The *Index of Economic Freedom* is published annually by the *Wall Street Journal* and The Heritage Foundation. Economic freedom is a crucial part of human development. Humans should have the right to move their belongings freely in order to move up, whether these are money, labour, knowledge, products or consumption. To the extent cultures and governments promote and protect freedom, they will enjoy more mobility.

The *Index of Economic Freedom* takes into account four pillars of economic freedom.*

1. Rule of law:
   a. Property rights: 'It measures the degree to which a country's laws protect private property rights and the extent to which those laws are respected. It also assesses the likelihood that private property will be expropriated by the state and analyses the independence of the judiciary, the existence of corruption within the judiciary, and the ability of individuals and businesses to enforce contracts.'

   b. Freedom from corruption: 'The score for this component is derived primarily from Transparency International's Corruption Perceptions Index (CPI), which measures the level of perceived corruption in 176 countries.'

2. Limited government:
   a. Fiscal Freedom: 'The fiscal freedom component is a composite measure of the burden of taxes that reflects both marginal tax rates and the overall level of taxation, including direct and indirect taxes imposed by all levels of government, as a percentage of GDP. The component score is derived from three quantitative factors:

* These notes were taken from the methodology of the *2014 Index of Economic Freedom* by The Heritage Foundation and the *Wall Street Journal*, and all quotations are from that report. Some conversion formulas and details are omitted, but can be checked at http://www.heritage.org/index/book/methodology

i. The top marginal tax rate on individual income,
ii. The top marginal tax rate on business income, and
iii. The total tax burden as a percentage of GDP.'

b. Government Spending: 'The Index methodology treats zero government spending as the benchmark. Underdeveloped countries, particularly those with little government capacity, may receive artificially high scores as a result. However, such governments, which can provide few if any public goods, are likely to receive low scores on some of the other components of economic freedom (such as property rights, financial freedom, and investment freedom) that measure aspects of government effectiveness. Government spending has a major impact on economic freedom, but it is just one of many important components. The scale for scoring government spending is non-linear, which means that government spending that is close to zero is lightly penalized, while levels of government spending that exceed 30 per cent of GDP lead to much worse scores in a quadratic fashion (for example, doubling spending yields four times less freedom). Only extraordinarily large levels of government spending – for example, over 58 per cent of GDP – receive a score of zero.' The formula for government expenditure is $GE_i = 100 - \alpha\ (\text{Expenditures}_i)^2$

3. Regulatory efficiency:
   a. Business Freedom: 'The business freedom score for each country is a number between 0 and 100, with 100 indicating the freest business environment. The score is based on 10 factors, all weighted equally, using data from the World Bank's Doing Business report.

i. Starting a business – procedures (number);
ii. Starting a business – time (days);
iii. Starting a business – cost (% of income per capita);
iv. Starting a business – minimum capital (% of income per capita);
v. Obtaining a licence – procedures (number);
vi. Obtaining a licence – time (days);
vii. Obtaining a licence – cost (% of income per capita);
viii. Closing a business – time (years);
ix. Closing a business – cost (% of estate); and
x. Closing a business – recovery rate (cents on the dollar).'

Each factor is converted to a scale of 0 to 100 using the following equation: $\text{Factor Score}_i = 50\ \text{factor}_{\text{average}} / \text{factor}_i$

b. Labour Freedom: The data also comes from the World Bank's Doing Business report, and it is converted with the same formula as the business freedom component. 'Six quantitative factors are equally weighted, with each counted as one-sixth of the labour freedom component:

i. Ratio of minimum wage to the average value added per worker,
ii. Hindrance to hiring additional workers,
iii. Rigidity of hours,
iv. Difficulty of firing redundant employees,
v. Legally mandated notice period, and
vi. Mandatory severance pay.'

c. Monetary Freedom: 'Monetary freedom combines a measure of price stability with an assessment of price controls. Both inflation and price controls distort market activity. Price stability without microeconomic intervention is the ideal state for the free market.

The score for the monetary freedom component is based on two factors:

i. The weighted average inflation rate for the most recent three years and
ii. Price controls.'

4. Open markets:
   a. Trade Freedom: 'Trade freedom is a composite measure of the extent of tariff and non-tariff barriers that affect imports and exports of goods and services. The trade freedom score is based on two inputs:

      i. The trade-weighted average tariff rate and
      ii. Non-tariff barriers (NTBs).'

   b. Investment Freedom. 'In an economically free country, there would be no constraints on the flow of investment capital. Individuals and firms would be allowed to move their resources into and out of specific activities, both internally and across the country's borders, without restriction. Such an ideal country would receive a score of 100 on the investment freedom component of the [Economic Freedom] Index.' Investment restrictions accounted in this part of the index include but are not limited to: national treatment of foreign investment, foreign investment code, restrictions on land ownership, sectorial investment restrictions, expropriation of investments without fair compensation, foreign exchange controls, capital controls.

c. Financial Freedom: 'Financial freedom is an indicator of banking efficiency as well as a measure of independence from government control and interference in the financial sector. State ownership of banks and other financial institutions such as insurers and capital markets reduces competition and generally lowers the level of access to credit . . . The Index scores an economy's financial freedom by looking into the following five broad areas:

   i. The extent of government regulation of financial services,
   ii. The degree of state intervention in banks and other financial firms through direct and indirect ownership,
   iii. The extent of financial and capital market development,
   iv. Government influence on the allocation of credit, and
   v. Openness to foreign competition.'

The Human Security Index (HSI) was recorded by a non-governmental organization, and is calculated for 232 countries. It is composed of three main elements:

- The Economic Fabric Index, which constitutes the economic dimension of security. It's about the financial security of both the individual and the nation.
- The Environmental Fabric Index, which represents some aspects of quality of life in terms of environmental factors.
- The Social Fabric Index, which represents the social aspects of security.

The variables considered to compute the Economic Fabric Index are:

- Income resources of a typical person, derived from:
  - GDP per capita at purchasing power parity – from the International Monetary Fund; World Development Indicators (WDI); Central Intelligence Agency (CIA).
  - Income equality (Gini coefficient) – from Solt, United Nations University World Institute for Development Economics Research (UNU-WIDER), WDI, CIA.
- Protection from financial catastrophe, derived from:
  - Foreign exchange reserves (percentage of imports) – from WDI, IMF, CIA.
  - External debt (as a percentage of GDP) – from WDI, CIA.
  - Current account balance (as a percentage of GDP) – from IMF, WDI, CIA.
- Health-care delivery/financing – digested from Carrin et al., 2001 and other sources.
- National savings rate (% of GDP per capita) – from the World Economic Forum, WDI, IMF, etc.

The following indicators were considered to calculate the Environmental Fabric Index:

- Environmental Vulnerability Index – from the Applied Geoscience and Technology Division of the Secretariat of the Pacific Community (SOPAC).
- Environmental Performance Index – from Yale University and the Centre for International Earth Science Information Network (CIESIN).

- Greenhouse gas emissions per capita – from the World Resources Institute and Wikipedia.
- Projected population growth rate 2010–2050 – from census.gov, UN Population Division, Statistics for Development of SOPAC.

The Social Fabric Index is comprised of the following three sub-components:

a) *Education and information empowerment*, which considers the next indicators:

   - Literacy rate (From UNESCO, WDI, CIA).
   - Connection Index, derived from:
     - Number of telephone lines per capita – International Telecommunication Union (ITU).
     - Number of mobile telephone accounts per capita – ITU.
     - Number of internet users per capita – ITU.
   - Press Freedom Index – from Reporters Without Borders.

b) *Diversity*: composed mainly of the Global Gender Gap Index from the World Economic Forum. Note that this gender gap is different from the Gender Inequality Index computed on the Sex component.

c) *Peacefulness*, which considers the following indicators:

   - Global Peace Index – from the Institute for Economics and Peace.
   - World Prison Population List/Brief – from the International Centre for Prison Studies.
   - Political Terror Scale – from politicalterrorscale.org

d) *Food security*, which considers the following indicators:

- Percentage of people undernourished – from the Food and Agriculture Organization, International Food Poverty Research Institute.
- Percentage of people below the local poverty index – from WDI, CIA.
- Food imports compared to exports and GDP – from WDI.
- Percentage of population experiencing food insecurity (needing emergency aid) – from the US Department of Agriculture.
- Percentage of productive land per capita, 2000+ – from WDI.
- Percentage change in productive land, 2000+ / 1960+ – derived from WDI.

e) *Health*, which considers the following indicators:

- Life expectancy at birth – from the World Health Organization (WHO), WDI, CIA.
- Percentage of life expectancy that is unhealthy – from WHO.
- Percentage of population using improved water source – from UNESCO, WDI.
- Health outcome equality – from Petrie and Tang (2008).

f) *Governance*, which considers the following indicators:

- Political stability, no violence – from World Governance Indicators (WGI).
- Control of illegal corruption – from WGI.
- Legal corruption – derived from WEF data, Vicente and Kaufmann (2011).

The rationale behind the HSI is that many challenges to human security are inherited by current decision-makers and leaders: natural hazards; trans-boundary threats or impacts; internal challenges from imperfect neighbourliness or governance; practices which in light of current situations may be harmful to people and their communities. So the Human Security Index aims to help people perceive and understand such situations, in order to strategize or to implement effective improvements to avoid them.

We chose the HSI since this indicator addressed more elements than the Human Development Index. We consider that the environment and the social fabric are important elements for moving up in a society, as important as the economic elements. The progress towards aspirations in people and communities is nothing less than moving up.

As we can see in the Security table, the results differ from the Human Development Index as the HSI considers elements of governance, sustainability and gender inequality, among others.

1.4 Success

For the Success component we used the World Economic Forum's Global Competitiveness Report for 2012–13. The report covers 143 countries.

The concept of competitiveness involves static and dynamic components. Although the productivity of a country determines its ability to sustain a high level of income, it is also one of the central determinants of its returns on investment, which is one of the key factors explaining an economy's growth potential. So, in this index, elements relating to competitiveness, rule of law and the importance of institutions inside a country (significant elements for success) are measured. It has twelve main pillars:

1. Institutions
2. Infrastructure
3. Macroeconomic environment
4. Health and primary education
5. Higher education and training
6. Goods and market efficiency
7. Labour market efficiency
8. Financial market development
9. Technological readiness
10. Market size
11. Business sophistication
12. Innovation

We considered this index when assessing the Success element of our index. The rationale is that as long as a country improves its efforts at developing competitiveness, it has more chance of achieving success in every way: in developing new business, in allowing internal competition in the country, in innovation, in knowledge, in improving the capacities and capabilities of the whole society.

As long as a country provides people with the tools and the elements to make them feel productive, successful, needed and part of its institutions and economy, people will surely be moving up. And, as long as people move up, a country will be moving up in terms of competitiveness, productivity and development.

Moving up is part of competition, productivity, availability, innovation and development. Moving up is all those things and more. However, even when a country fulfils those requirements, moving up is not guaranteed. It requires the combination of the other three S's.

We will now look in detail at the variables considered by the WEF for the twelve pillars that comprise their Global Competitiveness Report.

1.4.1. Institutions

• Property rights: Property rights and intellectual property protection.	• Security: Business costs of terrorism, business costs of crime and violence, organized crime and reliability of police services.
• Ethics and corruption: Diversion of public funds, public trust in politicians, and irregular payments and bribes.	• Government services for improved business performance. • Undue influence: Judicial independence and favouritism in decisions of government officials.
• Government efficiency: Wastefulness of government spending, burden of government regulation, efficiency of legal framework in challenging registers, efficiency of legal framework in settling disputes, transparency of government policymaking and provision of government services for improved business.	Private institutions: • Corporate ethics: Ethical behaviour of firms. • Accountability: Strength of auditing and reporting standards, protection of minority shareholders' interests, efficiency of corporate boards and strength of investor protection.

1.4.2. Infrastructure

Transport Infrastructure:	**Electricity and Telephony Infrastructure:**
• Quality of overall infrastructure. • Quality of railway infrastructure. • Quality of air transport infrastructure. • Available airline seats (km/week, millions). • Quality of roads.	• Fixed telephone lines/100 people. • Quality of electricity supply. • Mobile telephone subscriptions/100 people.

1.4.3. Macroeconomic Environment

- Government budget balance: percentage of GDP.
- Gross national savings: percentage of GDP.
- Inflation: annual percentage change.
- General government debt: percentage of the GDP.
- Country credit rating, 0–100 (100 is the best).

1.4.4. Health and Primary Education

Health:	Education:
• Business impact of malaria. • Malaria cases/100,000 people. • Business impact of tuberculosis. • Business impact of HIV/AIDS. • Life expectancy. • Infant mortality (deaths/1,000 live births). • Tuberculosis (cases/100,000 people). • HIV prevalence, percentage in adult population.	• Primary education enrolment (net percentage). • Quality of primary education.

1.4.5. Higher Education and Training

Quantity and Quality of Education:	On-the-job Training:
• Tertiary education enrolment (gross percentage).	• Local availability of research and training services. • Extent of staff training.

• Secondary education enrolment (gross percentage). • Quality of education system. • Quality of maths and science education. • Health access in schools.	

1.4.6. Goods Market Efficiency

Domestic Competition:	**Foreign Competition:**
• Intensity of local competition. • Extent of market dominance. • Extent and effect of taxation. • Effectiveness of anti-monopoly policy. • Total tax rate (percentage of profits). • Number of days required to start a business. • Number of procedures required to start a business. • Agricultural policy costs.	• Prevalence of trade barriers. • Trade tariffs (percentage paid as duty). • Prevalence of foreign ownership. • Business impact of rules on foreign direct investment (FDI). • Burden of customs regulations. • Imports as a percentage of GDP.

1.4.7. Labour Market Efficiency

Flexibility:	**Efficient use of talent:**
• Cooperation in employer–worker relations. • Flexibility of wage determination. • Hiring and firing practices. • Redundancy costs (weeks of salary). • Extent and effect of taxation.	• Pay and productivity. • Brain drain. • Reliance on professional management. • Women in labour force (ratio to men).

1.4.8. Financial Market Development

Efficiency:	**Trustworthiness and confidence:**
• Availability of financial services. • Affordability of financial services. • Financing through local equity market. • Ease of access to loans. • Venture capital availability.	• Soundness of banks. • Regulation of securities exchanges. • Legal rights index (0–10).

1.4.9. Technological Readiness

Technological adoption:	**Information and Communication Technology (ICT) use:**
• Availability of latest technologies. • Business-level technology absorption. • Foreign direct investment and technology transfer.	• Internet users. • Broadband Internet subscriptions. • Internet bandwidth. • Mobile broadband subscriptions. • Mobile telephone subscriptions. • Fixed telephone lines 1/2.

1.4.10. Market Size

- Domestic Market Size Index.
- Foreign Market Size Index.

1.4.11. Business Sophistication

- Local supplier quantity.
- Local supplier quality.
- State of cluster development.
- Nature of competitive advantage.
- Value chain breadth.
- Control of international distribution.

- Production process sophistication.
- Extent of marketing.
- Willingness to delegate authority.
- Reliance on professional management.

1.4.12. Innovation

- Capacity for innovation.
- Quality of scientific research institutions.
- Company spending on R&D.
- University–industry collaboration in R&D.
- Government procurement of advanced technology products.
- Availability of scientists and engineers.
- PCT patent applications.
- Intellectual property protection.

As we can see, there are multiple variables to determine competitiveness in a country. For our purpose, the more competitive a country is, the more upward mobility it will have in terms of our R^2 index.

1.5 The C^2: Evaluating the Cultural Unconscious

Dr Rapaille realized that, if we wanted to discover our unconscious and define the culture codes (C^2) of each country, then we would need to adopt a different approach from the traditional methodology with focus groups. So we decided to use 'The Rapaille Discovery Approach'.

The first step in this approach is that the questioner has to adopt the role of a 'professional stranger': a visitor from another planet who needs to convince people that he is a complete outsider who requires their help in understanding how a particular item or idea works and what emotions it is likely to provoke. Is money some kind of clothing? How does one appreciate love? Is status a sin? This allows people to begin the process of separating from their cortices and moving with the item or idea in question. What was the first emotion needed to produce the neurotransmitter in the brain to create and imprint the unconscious reference system, which is activated each time that they think about their own culture, country or nationality?

The sessions last three hours. The first hour is cortex, logical. We show participants a word, such as 'Chinese' or 'American', and we ask them to speak about it. As we do not believe what people say, we use this first hour as a purge, or 'washing-out session'. We usually do not discover anything new from the first hour, rather we retrieve the familiar clichés and stereotypes.

The second hour is more revealing, as we explore the latent structure which constitutes the culture. We use a method of word association that begins to reveal what is behind the cortex's intentions, for instance when Americans speak about freedom, and recall concepts such as 'the land of the free', 'freedom fighters' or 'freedom of speech'. We cannot really understand the meaning and the logic of emotion behind this word if we do not discover the latent structure. Just like the two profiles and the vase, we might just be aware of the profiles, when suddenly the vase reveals itself. In this example from American culture, one cannot comprehend the deeper meaning of American freedom without revealing the latent structure, or the vase, which in this case is prohibition. They are two sides of the same coin; two crucial elements of the same structure. One describes and defines the other. Therefore, in the land of the free you can expect sixteen years of alcohol prohibition, the dictatorship of political correctness (you cannot say 'black', you have to say 'African American'), not being allowed to smoke anywhere (even in New York parks), and no longer being able to buy an oversized Coke (now banned).

In French culture, this is a very different story. The latent structure for freedom, or *liberté*, is not prohibition ('*Oh, mon dieu, non!*'), rather it is privilege. The French want the freedom to kill those who are privileged in order to get this privilege for themselves. They are not against privilege, rather they are against other people having it. Napoleon, during his time as a revolutionary general, was against the nobility because they had privilege. Yet when he became emperor he re-established all the privileges for himself, his family and his friends.

By the third hour of the session – the point when the participants lie on the floor with pillows and listen to soothing music – people finally begin to say what they really mean. This process helps them access a different part of their brains. The answers they give now come from their reptilian brains, the place where instincts are housed. It is in our reptilian brains where the true answers are.

The third hour is the most interesting. After making them tell stories to a five-year-old coming from another planet (how weird is that?) during the second hour, we have managed to make the participants feel completely lost: they don't know what they are doing any more. This is exactly what we want, as they no longer know how to please us, and cannot understand what we are expecting from them. They are paid to come to the session, so they want to be good informants. Sometimes we can feel them silently begging, 'Just tell me what you want me to tell you', but this is not what we want, because this is the way they behave in most focus groups – they just try to guess what the leader wants them to say. This is why we don't believe what people say, instead we just ask them to lie down on the floor and go back to their very first imprint of what it means to be part of their culture.

We explain to them that we want them to remember, and in order to do that we want to re-create a mindset very similar to the one they have when they first wake up in the morning. Usually, when you wake, for the first five to ten minutes you can still remember your dreams, but if you don't write them down or record them immediately, you forget them and they are gone. This is because the cortex always arrives late to work in the morning.

When you wake up you are still emotionally involved, at the limbic level, in your dreams. You are also physically involved – for example, you might still be sweating if you were running away from something dangerous in your dream. Your heart might still be pounding. When your cortex arrives, it rationalizes the situation: 'This is just a dream, relax, there is no longer any danger.' It might also clean up your memory, and store everything away into your unconscious. You then quickly forget the dream and move on to what your conscious mind is asking you to do. This is the mindset we want to re-create: it helps people to go back and remember situations that they may have forgotten for decades.

We make it clear to the participants that this is going to be anonymous, they don't have to speak up if they don't want to. We provide them with pillows and blankets, they lie down on the floor and listen to a relaxation tape. After half an hour of relaxation, we take them back to their very first experience of the concept being studied – for instance, being Chinese or American. What was the first thing you can recall? How did you feel? Try to remember all of the details, the place, the colour, the smell, who was with you, what were the words that you used, what were the kinds of emotions and feelings that you experienced?

Then we ask them to go back to their most powerful experience associated with that concept, and finally we ask them to recall their most recent experience. We provide the participants with a notepad and paper so that they can record these experiences. They know that this is anonymous, and that they do not have to give out their name. We only ask them for their gender and age, to determine whether or not different structures exist within these criteria.

We collect all of these stories, totalling around 900 per culture, and look at the repetitions and patterns that exist within them. We don't study the content, but rather focus on the verbs to extract the structure. Do all of the stories have a common script? For example, if we look at *West Side Story* and *Romeo and Juliet*, or *Roxanne* and *Cyrano de Bergerac*, we realize that, though the content may be very different, the story is almost identical. The verbs have the same structure. Once we have discovered this pattern, or common structure, we open the file. The script is not new, it has existed in this culture for a long time. So we explore novels and movies, theatres and poetry, history and all expressions of that culture. We observe patterns and repetition.

Appendix 2: The Four S's in Numbers

$$R^2 = \frac{\text{Bio-Logical} + C^2}{2}$$

	Country	R^2	Bio-Logical	Rank Bio-Logical	C^2	Rank C^2
1	Switzerland	0.85	0.86	1	0.84	3
2	Canada	0.82	0.78	9	0.87	2
3	United States	0.81	0.74	14	0.88	1
4	Singapore	0.81	0.80	7	0.81	4
5	Germany	0.79	0.79	8	0.79	6
6	Norway	0.78	0.81	6	0.76	11
7	Australia	0.78	0.77	10	0.79	7
8	Finland	0.78	0.81	5	0.75	14
9	Austria	0.78	0.77	11	0.79	8
10	Denmark	0.77	0.82	3	0.72	15
11	New Zealand	0.77	0.77	12	0.77	10
12	Netherlands	0.76	0.82	4	0.70	17
13	Sweden	0.76	0.83	2	0.68	19
14	Israel	0.75	0.70	20	0.80	5
15	United Kingdom	0.75	0.73	16	0.76	12
16	Ireland	0.74	0.72	18	0.76	13
17	Korea	0.72	0.71	19	0.72	16
18	France	0.71	0.73	17	0.68	20
19	China	0.68	0.58	38	0.79	9
20	Chile	0.66	0.63	31	0.70	18
21	Japan	0.65	0.74	15	0.56	31
22	Belgium	0.64	0.76	13	0.52	33
23	Spain	0.63	0.67	21	0.59	28
24	Poland	0.63	0.63	28	0.63	24
25	Estonia	0.61	0.63	29	0.60	26
26	Brazil	0.61	0.54	43	0.68	21
27	Italy	0.61	0.66	23	0.55	32
28	Mexico	0.60	0.53	44	0.66	22
29	Czech Republic	0.58	0.67	22	0.50	36
30	Turkey	0.58	0.51	51	0.66	23
31	Malaysia	0.56	0.63	30	0.50	37
32	Portugal	0.56	0.66	24	0.47	38
33	Costa Rica	0.56	0.60	33	0.52	34
34	Colombia	0.56	0.51	50	0.61	25
35	Peru	0.55	0.51	48	0.60	27
36	Russia	0.53	0.48	57	0.59	29
37	Cyprus	0.53	0.65	26	0.41	47
38	Slovenia	0.52	0.63	27	0.41	48
39	Lithuania	0.52	0.60	35	0.44	43
40	Croatia	0.52	0.57	40	0.46	41
41	Latvia	0.51	0.60	36	0.43	44
42	Slovakia	0.51	0.60	32	0.42	45
43	Bulgaria	0.51	0.56	41	0.46	42
44	India	0.50	0.42	67	0.59	30
45	Hungary	0.50	0.59	37	0.41	49
46	Thailand	0.50	0.53	46	0.47	39
47	Indonesia	0.48	0.45	61	0.51	35
48	Kuwait	0.48	0.60	34	0.36	56

Appendix 2: The Four S's in Numbers

$$R^2 = \frac{\text{Bio-Logical} + C^2}{2}$$

	Country	R^2	Bio-Logical	Rank Bio-Logical	C^2	Rank C^2
49	Ukraine	0.48	0.48	56	0.47	40
50	Greece	0.47	0.57	39	0.37	54
51	Argentina	0.46	0.50	52	0.42	46
52	Romania	0.45	0.52	47	0.39	52
53	Vietnam	0.45	0.49	55	0.41	50
54	South Africa	0.45	0.50	53	0.39	53
55	Morocco	0.43	0.46	59	0.40	51
56	El Salvador	0.41	0.44	62	0.37	55
57	Saudi Arabia	0.38	0.55	42	0.22	59
58	UAE	0.37	0.66	25	0.09	67
59	Philippines	0.37	0.46	58	0.28	57
60	Jordan	0.36	0.50	54	0.22	60
61	Kazakhstan	0.35	0.51	49	0.19	62
62	Dominican Republic	0.33	0.43	65	0.23	58
63	Iran	0.29	0.44	64	0.14	63
64	Azerbaijan	0.28	0.53	45	0.04	71
65	Kenya	0.28	0.35	69	0.21	61
66	Ecuador	0.28	0.43	66	0.13	64
67	Algeria	0.28	0.44	63	0.11	66
68	Sri Lanka	0.27	0.46	60	0.09	68
69	Venezuela	0.23	0.40	68	0.06	70
70	Pakistan	0.23	0.33	71	0.13	65
71	Bangladesh	0.22	0.34	70	0.09	69

Appendix 2: The Four S's in Numbers

$$\textbf{Bio-Logical} = \frac{\text{Survival} + \text{Sex} + \text{Security} + \text{Success}}{4}$$

	Country	Bio-Logical	Survival	Sex	Security	Success	Graph
1	Switzerland	**0.86**	0.73	0.93	0.81	0.98	Survival 0.73 Security 0.81 Success 0.98 Sex 0.93
2	Sweden	**0.83**	0.69	0.95	0.78	0.92	Survival 0.69 Security 0.78 Success 0.92 Sex 0.95
3	Denmark	**0.82**	0.74	0.94	0.77	0.85	Survival 0.74 Security 0.77 Success 0.85 Sex 0.94
4	Netherlands	**0.82**	0.68	0.95	0.74	0.91	Survival 0.68 Security 0.74 Success 0.91 Sex 0.95
5	Finland	**0.81**	0.63	0.93	0.77	0.92	Survival 0.63 Security 0.77 Success 0.92 Sex 0.93
6	Norway	**0.81**	0.69	0.92	0.78	0.84	Survival 0.69 Security 0.78 Success 0.84 Sex 0.92
7	Singapore	**0.80**	0.53	0.91	0.81	0.96	Survival 0.53 Security 0.81 Success 0.96 Sex 0.91
8	Germany	**0.79**	0.59	0.91	0.74	0.90	Survival 0.59 Security 0.74 Success 0.90 Sex 0.91

Appendix 2: The Four S's in Numbers

$$\textbf{Bio-Logical} = \frac{\text{Survival} + \text{Sex} + \text{Security} + \text{Success}}{4}$$

	Country	Bio-Logical	Survival	Sex	Security	Success	Graph
9	Canada	**0.78**	0.65	0.86	0.76	0.84	Survival 0.65 Security 0.76 Success 0.84 Sex 0.86
10	Australia	**0.77**	0.65	0.86	0.78	0.79	Survival 0.65 Security 0.78 Success 0.79 Sex 0.86
11	Austria	**0.77**	0.65	0.87	0.74	0.82	Survival 0.65 Security 0.74 Success 0.82 Sex 0.87
12	New Zealand	**0.82**	0.71	0.80	0.78	0.78	Survival 0.71 Security 0.78 Success 0.78 Sex 0.80
13	Belgium	**0.81**	0.63	0.89	0.71	0.82	Survival 0.63 Security 0.71 Success 0.82 Sex 0.89
14	United States	**0.81**	0.69	0.70	0.67	0.90	Survival 0.69 Security 0.67 Success 0.90 Sex 0.70
15	Japan	**0.80**	0.48	0.88	0.73	0.88	Survival 0.48 Security 0.73 Success 0.88 Sex 0.88
16	United Kingdom	**0.79**	0.51	0.79	0.73	0.89	Survival 0.51 Security 0.73 Success 0.89 Sex 0.79

Appendix 2: The Four S's in Numbers

$$\textbf{Bio-Logical} = \frac{\text{Survival} + \text{Sex} + \text{Security} + \text{Success}}{4}$$

	Country	Bio-Logical	Survival	Sex	Security	Success	Graph
17	France	**0.73**	0.55	0.89	0.68	0.79	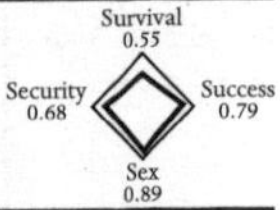 Survival 0.55; Security 0.68; Success 0.79; Sex 0.89
18	Ireland	**0.72**	0.63	0.80	0.73	0.73	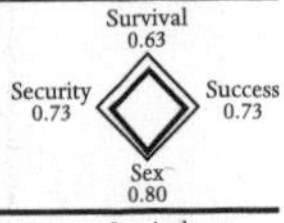 Survival 0.63; Security 0.73; Success 0.73; Sex 0.80
19	Korea	**0.71**	0.49	0.89	0.67	0.79	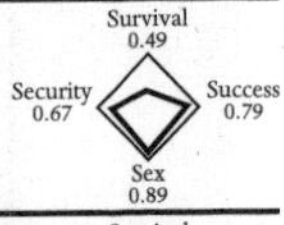 Survival 0.49; Security 0.67; Success 0.79; Sex 0.89
20	Israel	**0.70**	0.53	0.86	0.66	0.76	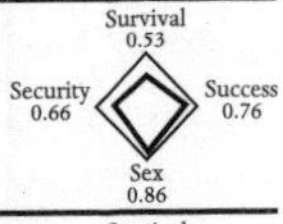 Survival 0.53; Security 0.66; Success 0.76; Sex 0.86
21	Spain	**0.67**	0.49	0.88	0.69	0.64	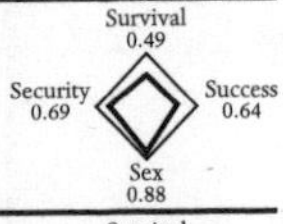 Survival 0.49; Security 0.69; Success 0.64; Sex 0.88
22	Czech Republic	**0.67**	0.45	0.86	0.74	0.61	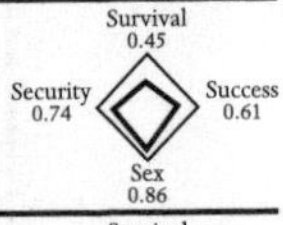 Survival 0.45; Security 0.74; Success 0.61; Sex 0.86
23	Italy	**0.66**	0.53	0.88	0.66	0.59	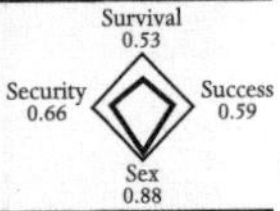 Survival 0.53; Security 0.66; Success 0.59; Sex 0.88
24	Portugal	**0.66**	0.52	0.86	0.67	0.58	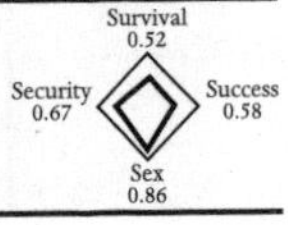 Survival 0.52; Security 0.67; Success 0.58; Sex 0.86

Appendix 2: The Four S's in Numbers

$$\textbf{Bio-Logical} = \frac{\text{Survival} + \text{Sex} + \text{Security} + \text{Success}}{4}$$

	Country	Bio-Logical	Survival	Sex	Security	Success	Graph
25	United Arab Emirates	**0.66**	0.42	0.77	0.66	0.78	Survival 0.42 Security 0.66 Success 0.78 Sex 0.77
26	Cyprus	**0.65**	0.51	0.86	0.69	0.55	Survival 0.51 Security 0.69 Success 0.55 Sex 0.86
27	Slovenia	**0.63**	0.47	0.82	0.68	0.56	Survival 0.47 Security 0.68 Success 0.56 Sex 0.82
28	Poland	**0.63**	0.40	0.84	0.69	0.59	Survival 0.40 Security 0.69 Success 0.59 Sex 0.84
29	Estonia	**0.63**	0.32	0.81	0.74	0.65	Survival 0.32 Security 0.74 Success 0.65 Sex 0.81
30	Malaysia	**0.63**	0.35	0.71	0.68	0.78	Survival 0.35 Security 0.68 Success 0.78 Sex 0.71
31	Chile	**0.63**	0.49	0.63	0.74	0.65	Survival 0.49 Security 0.74 Success 0.65 Sex 0.63
32	Slovakia	**0.60**	0.41	0.81	0.70	0.50	Survival 0.41 Security 0.70 Success 0.50 Sex 0.81

Appendix 2: The Four S's in Numbers

$$\textbf{Bio-Logical} = \frac{\text{Survival} + \text{Sex} + \text{Security} + \text{Success}}{4}$$

	Country	Bio-Logical	Survival	Sex	Security	Success	Graph
33	Costa Rica	**0.60**	0.54	0.64	0.67	0.56	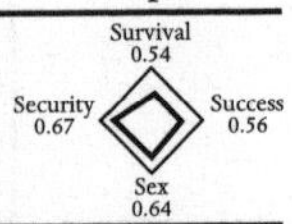 Survival 0.54; Security 0.67; Success 0.56; Sex 0.64
34	Kuwait	**0.60**	0.39	0.77	0.62	0.62	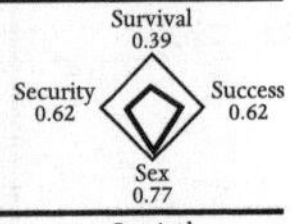 Survival 0.39; Security 0.62; Success 0.62; Sex 0.77
35	Lithuania	**0.60**	0.28	0.81	0.73	0.58	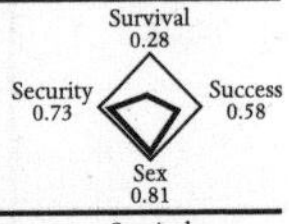 Survival 0.28; Security 0.73; Success 0.58; Sex 0.81
36	Latvia	**0.60**	0.33	0.78	0.71	0.56	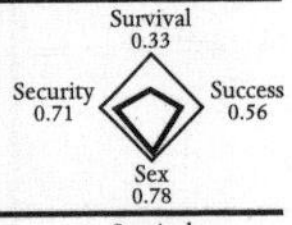 Survival 0.33; Security 0.71; Success 0.56; Sex 0.78
37	Hungary	**0.59**	0.36	0.76	0.70	0.55	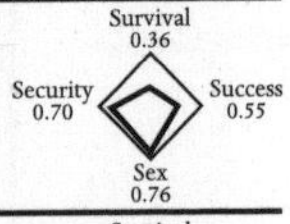 Survival 0.36; Security 0.70; Success 0.55; Sex 0.76
38	China	**0.58**	0.22	0.79	0.58	0.72	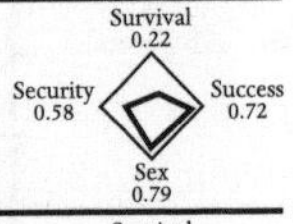 Survival 0.22; Security 0.58; Success 0.72; Sex 0.79
39	Greece	**0.57**	0.43	0.84	0.61	0.41	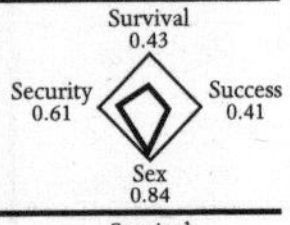 Survival 0.43; Security 0.61; Success 0.41; Sex 0.84
40	Croatia	**0.57**	0.32	0.83	0.66	0.47	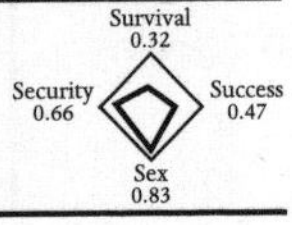 Survival 0.32; Security 0.66; Success 0.47; Sex 0.83

Appendix 2: The Four S's in Numbers

$$\textbf{Bio-Logical} = \frac{\text{Survival} + \text{Sex} + \text{Security} + \text{Success}}{4}$$

	Country	Bio-Logical	Survival	Sex	Security	Success	Graph
41	Bulgaria	**0.56**	0.26	0.75	0.68	0.54	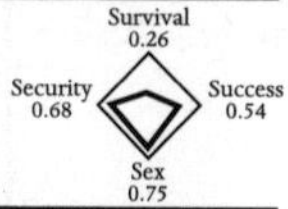 Survival 0.26; Security 0.68; Success 0.54; Sex 0.75
42	Saudi Arabia	**0.55**	0.39	0.35	0.63	0.82	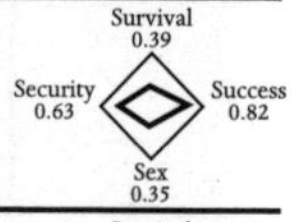 Survival 0.39; Security 0.63; Success 0.82; Sex 0.35
43	Brazil	**0.54**	0.43	0.55	0.59	0.58	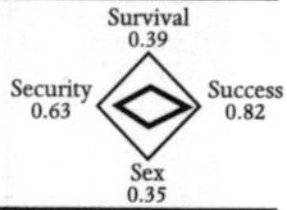 Survival 0.39; Security 0.63; Success 0.82; Sex 0.35
44	Mexico	**0.53**	0.36	0.55	0.64	0.56	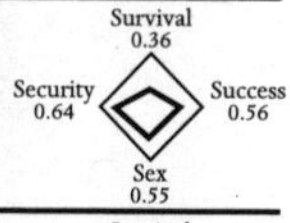 Survival 0.36; Security 0.64; Success 0.56; Sex 0.55
45	Azerbaijan	**0.53**	0.21	0.69	0.64	0.58	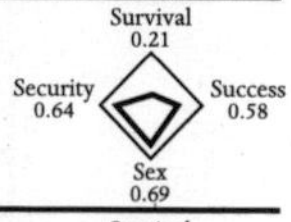 Survival 0.21; Security 0.64; Success 0.58; Sex 0.69
46	Thailand	**0.53**	0.24	0.62	0.64	0.61	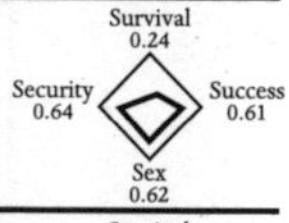 Survival 0.24; Security 0.64; Success 0.61; Sex 0.62
47	Romania	**0.52**	0.24	0.67	0.68	0.48	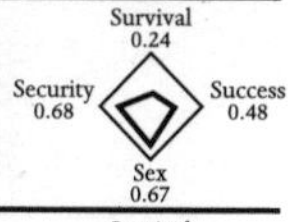 Survival 0.24; Security 0.68; Success 0.48; Sex 0.67
48	Peru	**0.51**	0.26	0.58	0.65	0.54	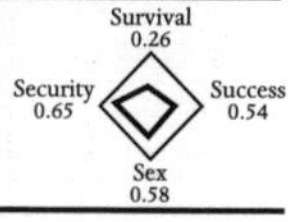 Survival 0.26; Security 0.65; Success 0.54; Sex 0.58

Appendix 2: The Four S's in Numbers

$$\textbf{Bio-Logical} = \frac{\text{Survival} + \text{Sex} + \text{Security} + \text{Success}}{4}$$

	Country	Bio-Logical	Survival	Sex	Security	Success	Graph
49	Kazakhstan	**0.51**	0.14	0.67	0.66	0.57	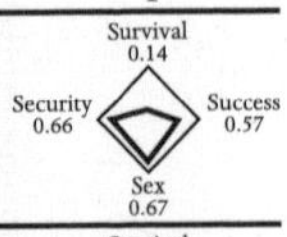 Survival 0.14 Security 0.66 Success 0.57 Sex 0.67
50	Colombia	**0.51**	0.34	0.52	0.66	0.51	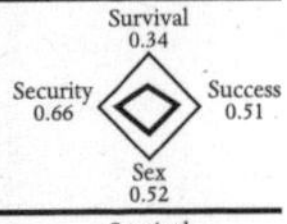 Survival 0.34 Security 0.66 Success 0.51 Sex 0.52
51	Turkey	**0.51**	0.26	0.56	0.62	0.59	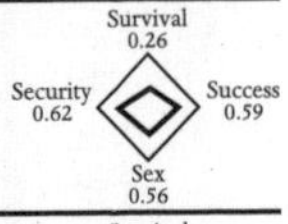 Survival 0.26 Security 0.62 Success 0.59 Sex 0.56
52	Argentina	**0.50**	0.41	0.63	0.56	0.42	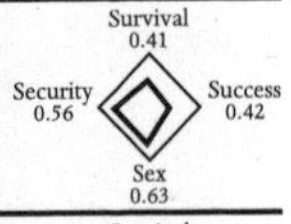 Survival 0.41 Security 0.56 Success 0.42 Sex 0.63
53	South Africa	**0.50**	0.34	0.51	0.59	0.57	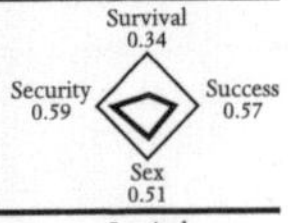 Survival 0.34 Security 0.59 Success 0.57 Sex 0.51
54	Jordan	**0.50**	0.29	0.54	0.65	0.52	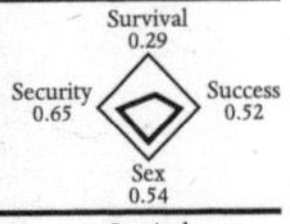 Survival 0.29 Security 0.65 Success 0.52 Sex 0.54
55	Viet Nam	**0.49**	0.21	0.69	0.55	0.49	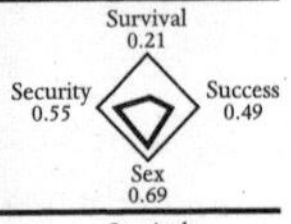 Survival 0.21 Security 0.55 Success 0.49 Sex 0.69
56	Ukraine	**0.48**	0.19	0.66	0.57	0.50	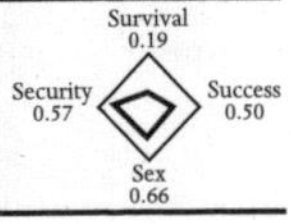 Survival 0.19 Security 0.57 Success 0.50 Sex 0.66

Appendix 2: The Four S's in Numbers

$$\textbf{Bio-Logical} = \frac{\text{Survival} + \text{Sex} + \text{Security} + \text{Success}}{4}$$

	Country	Bio-Logical	Survival	Sex	Security	Success	Graph
57	Russia	**0.48**	0.15	0.66	0.58	0.52	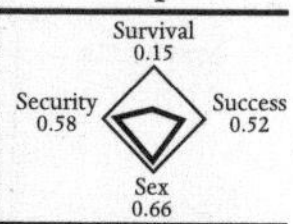 Survival 0.15; Security 0.58; Success 0.52; Sex 0.66
58	Philippines	**0.46**	0.16	0.57	0.57	0.52	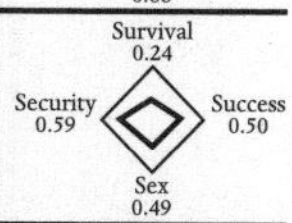 Survival 0.24; Security 0.59; Success 0.50; Sex 0.49
59	Morocco	**0.46**	0.24	0.49	0.59	0.50	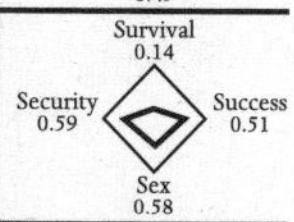 Survival 0.14; Security 0.59; Success 0.51; Sex 0.58
60	Sri Lanka	**0.46**	0.14	0.58	0.59	0.51	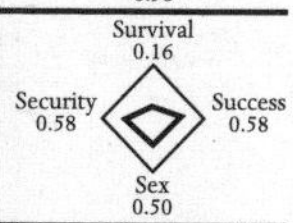 Survival 0.16; Security 0.58; Success 0.58; Sex 0.50
61	Indonesia	**0.45**	0.16	0.50	0.58	0.58	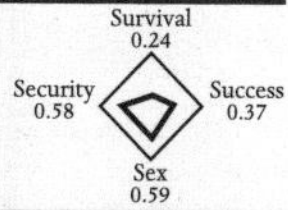 Survival 0.24; Security 0.58; Success 0.37; Sex 0.59
62	El Salvador	**0.44**	0.23	0.51	0.64	0.39	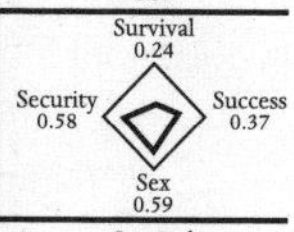 Survival 0.24; Security 0.58; Success 0.37; Sex 0.59
63	Algeria	**0.44**	0.24	0.59	0.58	0.37	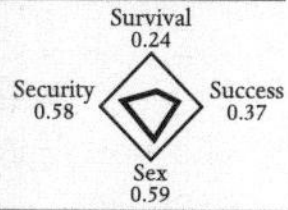 Survival 0.24; Security 0.58; Success 0.37; Sex 0.59
64	Iran	**0.44**	0.24	0.52	0.49	0.52	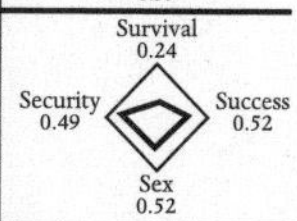 Survival 0.24; Security 0.49; Success 0.52; Sex 0.52

Appendix 2: The Four S's in Numbers

$$\textbf{Bio-Logical} = \frac{\text{Survival} + \text{Sex} + \text{Security} + \text{Success}}{4}$$

	Country	Bio-Logical	Survival	Sex	Security	Success	Graph
65	Dominican Republic	**0.43**	0.23	0.52	0.59	0.38	Survival 0.23; Security 0.59; Success 0.38; Sex 0.52
66	Philippines	**0.43**	0.21	0.53	0.53	0.44	Survival 0.21; Security 0.53; Success 0.44; Sex 0.53
67	Morocco	**0.42**	0.18	0.38	0.56	0.55	Survival 0.18; Security 0.56; Success 0.55; Sex 0.38
68	Sri Lanka	**0.40**	0.26	0.55	0.48	0.29	Survival 0.26; Security 0.48; Success 0.29; Sex 0.55
69	Indonesia	**0.35**	0.14	0.37	0.52	0.38	Survival 0.14; Security 0.52; Success 0.38; Sex 0.37
70	El Salvador	**0.34**	0.05	0.45	0.52	0.35	Survival 0.05; Security 0.52; Success 0.35; Sex 0.45
71	Algeria	**0.33**	0.04	0.43	0.53	0.31	Survival 0.04; Security 0.53; Success 0.31; Sex 0.43

ILLUSTRATION CREDITS

Grateful acknowledgement is given to the following for permission to reproduce copyrighted material:

p. xiv: Anthony Taber/The New Yorker Collection/The Cartoon Bank; p. 30: Danny Shanahan/The New Yorker Collection/The Cartoon Bank; p. 40: www.CartoonStock.com; p. 58: www.CartoonStock.com; p. 84: Elmgreen & Dragset, *Social Mobility*, installation shot from Bergen Kunsthall, 2005, courtesy the artists (photograph: Thor Brødreskift); p. 91: Dana Fradon/The New Yorker Collection/The Cartoon Bank; p. 80: copyright © Steve Leard, 2009; p. 94: Michael Crawford/The New Yorker Collection/The Cartoon Bank; p. 95: www.CartoonStock.com; p. 101: Lee Lorenz/The New Yorker Collection/The Cartoon Bank; p. 117: Peter C. Vey/The New Yorker Collection/The Cartoon Bank; p. 146: www.CartoonStock.com; p. 166: Henry Martin/The New Yorker Collection/The Cartoon Bank; p. 175: Roz Chast/The New Yorker Collection/The Cartoon Bank; p. 179: www.CartoonStock.com; p. 210: courtesy of the Leo Baeck Institute.